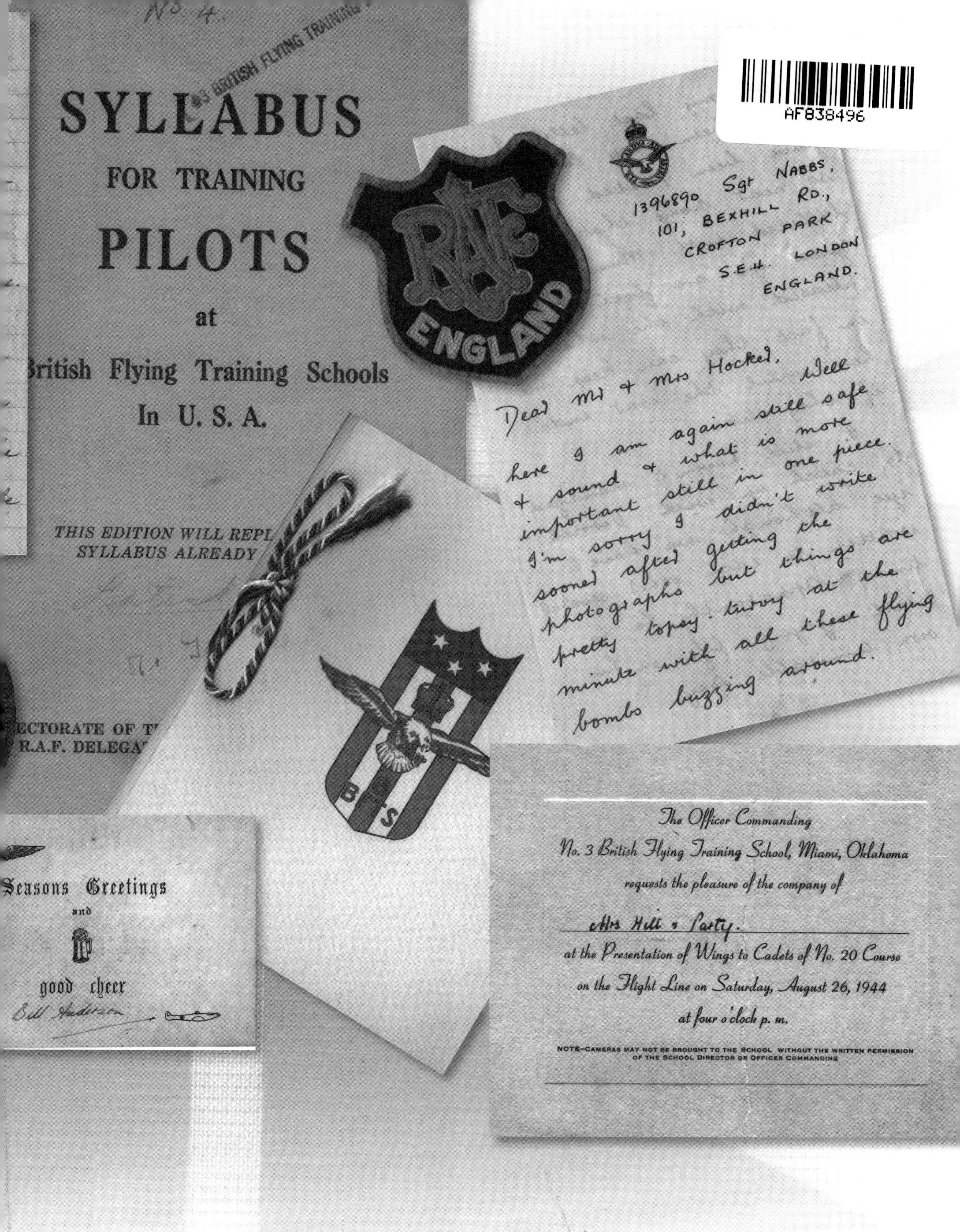
No. 4.
#3 BRITISH FLYING TRAINING
SYLLABUS
FOR TRAINING
PILOTS
at
British Flying Training Schools
In U. S. A.
THIS EDITION WILL REPL
SYLLABUS ALREADY
ECTORATE OF T
R.A.F. DELEGA
RAF
ENGLAND
1396890 Sgt NABBS,
101, BEXHILL RD.,
CROFTON PARK
S.E.4. LONDON
ENGLAND.
Dear Mr & Mrs Hocked, Well
here I am again still safe
& sound & what is more
important still in one piece.
I'm sorry I didn't write
sooner after getting the
photographs but things are
pretty topsy-turvy at the
minute with all these flying
bombs buzzing around.
BFTS
Seasons Greetings
and
good cheer
Bill Anderson
The Officer Commanding
No. 3 British Flying Training School, Miami, Oklahoma
requests the pleasure of the company of
Mrs Hill & Party.
at the Presentation of Wings to Cadets of No. 20 Course
on the Flight Line on Saturday, August 26, 1944
at four o'clock p. m.
NOTE—CAMERAS MAY NOT BE BROUGHT TO THE SCHOOL WITHOUT THE WRITTEN PERMISSION OF THE SCHOOL DIRECTOR OR OFFICER COMMANDING

The Royal Air Force in Oklahoma

Loves, Lives & Courage of the British Air Crews Trained in Oklahoma During World War II

FUNDING MADE POSSIBLE BY

Mr. & Mrs. Jack C. Bowker

George Bradford Butcher

Leslie Butcher

Daughters of the American Revolution, Ponca City Chapter

Mr. & Mrs. Keith R. Hunter

Mathews Children's Trust

No. 6 British Flying Training School Association

North Central Oklahoma Historical Association, Inc.

Ponca City Publishing Company, Inc.

Paul H. & Zeta D. Powell

Steven & Sharon Primeau

Lucie Wright

The Royal Air Force in Oklahoma

Loves, Lives & Courage of the British Air Crews Trained in Oklahoma During World War II

BY PAULA CARMACK DENSON

SERIES EDITOR: GINI MOORE CAMPBELL

Oklahoma Horizons Series

OKLAHOMA HERITAGE ASSOCIATION
OKLAHOMA CITY

The Royal Air Force in Oklahoma

Loves, Lives & Courage of the British Air Crews Trained in Oklahoma During World War II

Library of Congress Control Number: 2006935831

ISBN 978-1-885596-56-7

Printed in the United States of America

Jacket Design by Sandi Welch

Book Design by Kris Vculek

OKLAHOMA HERITAGE ASSOCIATION
OKLAHOMA CITY, OKLAHOMA
1.888.501.2059
www.oklahomaheritage.com

CONTENTS

Three Royal Air Force cadets stationed at Darr School of Aeronautics.

Courtesy John Cal Price

— Acknowledgments —

Long before I ever dreamed of actually writing this book I began collecting bits and pieces of information about the two Oklahoma British Flying Training Schools. I owe many thanks to Mary Ellen Cantrell, who initially became involved in planting the idea of writing a book in my head, and for all the people who provided me with information and photographs as I began my research.

Without Lillian Taylor, my Ponca City mentor, who shared a great deal of her knowledge with me in the initial stages of my research, I would never been able to start this book and, of course, I am immensely grateful to the trustees of the Oklahoma Heritage Association for the support which they have so generously given in seeing that this book became a reality. In particular, I want to thank Gini Moore Campbell, who has believed in this project from the start and has provided me with splendid editorial assistance. Without her help and the talents of Kris Vculek, Sandi Welch, and Nancy Peterson, the preservation of this valuable part of Oklahoma history could not have happened.

I wish to thank the faculty and staff of Northeastern Oklahoma A&M College in Miami. Without the help of Jim Genandt, Sloane Brown, Alberta Hutchings, and Edie Ingram, I never would have been able to learn so much about the Spartan School of Aeronautics and No. 3 BFTS. Judy Johnson, manager of the Dobson Museum in Miami, was also very generous with her time and records. There were others in Miami who lent a hand, including motel clerks, Jane and Willie Osborn of the Coleman Theatre, the Miami Chamber of Commerce, and the Miami Public Library.

The Ponca City Library staff, including Sandy Graves and Loyd Bishop, deserves a big round of applause. My friend, Bret Carter, has consistently and unselfishly shared some of his amazing photographs and I certainly owe him a big debt of thanks. Reaching farther out of the Ponca City community, I also received special help from No. 6 British Flying Training School Ponca City Association, and Eddy Roach of *The Daily Oklahoman*. Jim Dawson and Tod Roberts provided helpful assistance with suggestions for my research as well as Mrs. Jill Prime, Sandy Trotter, John Hays, Barbara Davis, Bill Moore, Mr. and Mrs. Jim Hocker, Jean and Mark Sealey, Marjorie Shelton, Derek Inskeep, John O'Brien, Baroness Detta O'Cathain, a member of the British Parliament, and Wing Commander Mike Crawley, the RAF British Liaison in Washington, D.C. The staff of the Imperial War Museum and Gordon Leith, director of the Hendon Museum in London, England, were also very generous with their time and help. Others I need to thank are the Anthony, Kansas library staff and the Anthony, Kansas Chamber of Commerce.

In my own hometown of Ponca City, I have also received generous support and help from Karen and Keith Hunter, Mr. and Mrs. Fred Clarke, Mr. and Mrs. Jack Bowker, Mr. and Mrs. Tom Muchmore of the Ponca City Publishing Company, the Ponca City Chapter of the Daughters of the American Revolution, Friends of Marland's Grand Home Foundation, and the North Central Oklahoma Historical Association. Dr. and Mrs. J.E. Fuller, Mr. and Mrs. J.E. Ledbetter, Mr. Jean O'Kelley, Mrs. Hugh Wiles, Bill Price, Kristin Weltge Webb, Mrs. Robert K. Smith, Mrs. Johanne Mead, Mrs. Harold Cogman, Dr. C. Thomas Hill, Mr. and Mrs. Preston Gant, Mrs. Betty St. Clair, and Mrs. Mariann Bryant have also shared their knowledge and enthusiasm in this project.

Former Ponca City friends including Steven and Sharon Primeau, Sarah Barrington Westerheide, and Paul and Zeta Powell also have helped to make this dream come true, along with other Oklahomans including Leslie Butcher and Lucie Wright. Thank you!

Thanks and love to my family, including Sheryl Sallee, Peggy Park, and Patti Carmack. My late mother, Erma Burleson Carmack, also shared her memories of the cadets in Ponca City. Most especially, I owe a great debt of thanks to my husband, Robert, and my daughter and son-in-law, Mr. and Mrs. Anthony Goebel, for their kind and loving support all along this incredible journey.

Finally, I could not have accomplished this book without special friends including former cadets and family members of both No. 6 BFTS and No. 3 BFTS. Most helpful from No. 6 BFTS were John Cal Price, Albert "Bert" Sims, Dennis Stone, Mike Igglesden, Roland Whittaker, Ted Newman, with help from his son David, Arthur Such, Brian Barnes, Mr. and Mrs. David Olorenshaw, Albert Spencer, Tony Norton, Mr. and Mrs. Andrew Coleby, Ken Westcott, Noel Clark, Air Vice-Marshal H. Gill, Robert Kelson, son of Mr. William B. Kelson, Mr. and Mrs. David Riekie, Mr. and Mrs. Jerry Eaton, Mr. and Mrs. Richard Ravitts, John Chadwick, William Williams, Ian and Owen Salkin, Robert Aston, Kenneth Bellamy, Jack Dalton, Mike Dobson, Mrs. Cindy Stevens, Ronald Wakeman, Stan Ward, Eric Hookings, Reg Gilbert, Alan Stredwick, Rick Brown, Vernon Radford, Peter Greaves, Bert Ginsberg, Eddy Pugh, Malcolm Norman, Jack Youens, and Mr. William Neilson.

Those cadets and families from No. 3 BFTS are also owed a huge debt of gratitude. They are Mr. and Mrs. Deryk Maddox, Alan Thomas, Stanley Duck, Raymond Baxter, N.N. "Nick" Brown, Dave Davis, David Dinnie, Doug Eaton, Ron Emmerson, John Facer, Cyril Grogan, David Hammond, Brian Haslam, J.F. Langer, A.R. Marsh, Ian McDonald, Roger Mills, Ken Odell, John Quine, M.H. Thirkettle, William J. West, Tony and Adrian Wadley, and Kay Wessale Schnieber. I also wish to thank the others who helped me but are nameless. Although I carefully tried to record all names, I am certain that there are a few I have forgotten or did not record. Please be assured this is an unintentional error on my part and that these contributions also are gratefully acknowledged.

Barn Hill
Hunton
Maidstone
Aug. 24. 8. 44. Kent.

Dear Mrs Hocker

Many thanks for your kind letter we was very pleased to have a letter but sorry you have felt worried about Dick through his not writing I know he as been very busy the last few months as Violet as been writing to us as Dick had not the time.

We had a letter from Violet this morning saying they was well Dick had been away to another place but was back again they are away from the Flying bombs but they come over us to get to London they are horrid things we have seen the Fighters bring a lot down & I am afraid they have caused a lot of life & damage.

— Preface —

Paula Carmack Denson

As a native of Ponca City, I was born in a town rich in history, including the establishment and operation of Darr School of Aeronautics, Ponca City's Royal Air Force British Flying Training School, which had closed just one year before I was born. However, as a child and as a young adult, I was too busy pursuing my own youthful interests to realize what a treasure-trove of history my state and local surroundings contained, including the founding of two extraordinary Royal Air Force schools in Oklahoma. However, as I matured, I became addicted to seeking facts and photos about events which had occurred in my "neck of the woods." Because of this passion, I was responsible for founding North Central Oklahoma Historical Association in 1994, a non-profit organization dedicated to preserving the history of my county. Together, with a board of dedicated directors, we successfully published a national award winning two-volume history titled *North Central Oklahoma: Rooted in the Past, Growing for the Future*. This set the stage for further research and publication goals.

Thinking back, perhaps I heard enough flying tales as I grew up to subliminally pique my interest in aviation. Judging from the picture of me as a two year old, "flying" down the street with my first cousin, James E. Ledbetter, I may have always taken some interest in matters of this sort. So must have my cousin, for he joined the United States Air Force as a young man, eventually achieving the rank of Colonel.

My interest in the history of these schools occurred much later and quite by accident. I suppose it was blind luck that I attended an estate auction in 1997 and came away with a two-dollar shoebox full of old letters. They were postmarked from various locations in Great Britain during the years 1942 through 1947 and I was curious about their contents because that was during the years surrounding my birth. It was only after I sifted through the letters at a later time that I became aware that most were written by men who trained at Darr School of Aeronautics. These epistles were written to the Hocker family, one of many families who hosted the British while they were in Ponca City. I decided to learn more about the authors of these letters and I began a search which led me to some of the men who penned the letters. I have met some of them and their families and, as a result, I have established friendships all over the world. What a thrill to actually meet some of my heroes! With the generous help of many of them, I have become addicted to learning more and more about their involvement with Oklahoma.

Recently, my search was expanded to include learning more about the northeastern Oklahoma town of Miami, Oklahoma, and its role in hosting No. 3 BFTS at Spartan School of Aeronautics. Eventually, with the help of many talented and friendly people, I have been launched into a publishing project which seems especially appropriate as Oklahomans celebrate their centennial. The time is right for the story of these brave young men to come to light and so I leave it with you to enjoy. As my British friends would say – "Tails Up"!

– 2006

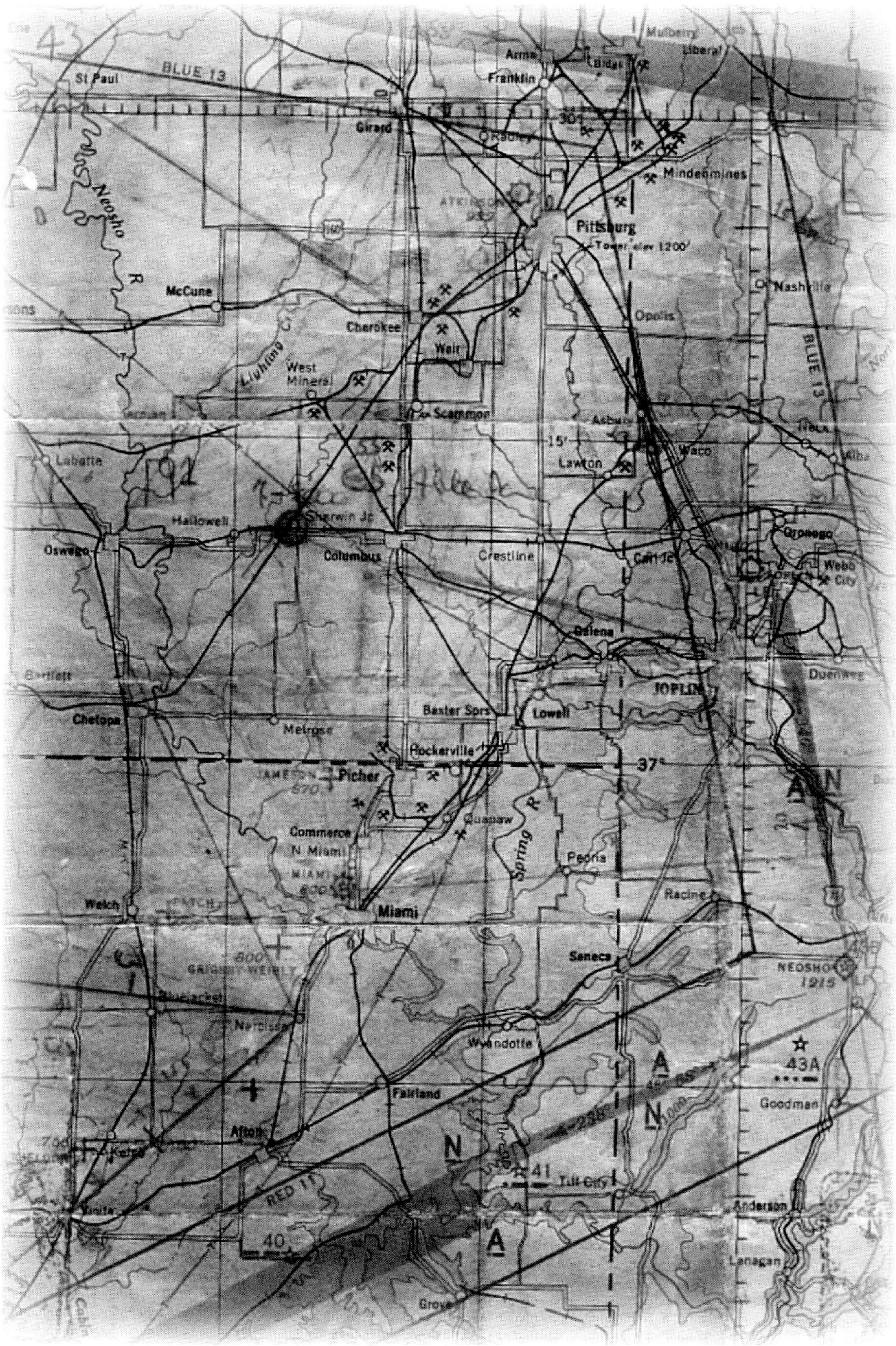

Navigational maps of the area where training would occur were sometimes printed on silk scarves, so that if the men were lost they would have some help close at hand. *Courtesy John Facer*

— Dedication —

High Flight

Oh! I have slipped the surly bonds of earth
And danced the skies on laughter-silvered wings;
Sunward I've climbed, and joined the tumbling mirth
Of sun-split clouds - and done a hundred things
You have not dreamed of – wheeled and soared and swung
High in the sunlit silence. Hov'ring there,
I've chased the shouting wind along, and flung
My eager craft through footless halls of air.

Up, up the long, delirious burning blue
I've topped the wind-swept heights with easy grace
Where never lark, nor even eagle flew—
And, while with silent lifting mind I've trod
The high, un-trespassed sanctity of space,
Put out my hand and touched the face of God.

Pilot Office John Gillespie Magee, Jr.
No. 412 Squadron, RCAF
Killed December 11, 1941

This book is dedicated to the cadets of No. 6 BFTS and No. 3 BFTS and to all who served them while they learned to fly over Oklahoma prairies. It is also dedicated to the many families who adopted these young men into their lives and who have shared their stories and photographs.

Most especially, it is dedicated to my husband, Robert, for his loving and unselfish patience while I spent my time in many other places in order to see this book created.

PCD

Three Ponca City RAF cadets in uniform.

Courtesy Betty Page St. Clair

— Foreword —

Raymond Baxter, OBE

One of the most remarkable and rewarding examples of Anglo-American cooperation during World War II was the establishment of the British Flying Training Schools in the United States. It is important to remember that this was before Pearl Harbor forced the United States into the War and while there were powerful voices in government and public opinion which were vociferously opposed to America being "dragged into another European War". An example of this tense state of affairs, as reported in this book, was that when my colleagues and I traveled from Moncton, New Brunswick, to Miami, Oklahoma, to join No. 5 course at 3BFTS we wore civilian clothes as had our predecessors. The fact that we were in a special train and that our grey flannel suits, blue shirts and ties were identical was presumably in the hope that we might pass unnoticed by those who may have objected to the presence of a train-load of uniformed foreign troops on American soil.

Indeed, there were a few displays of open hostility in the early days but these were dwarfed into insignificance by the generous and warm-hearted welcome of the vast majority of our American hosts. It was for them, as for us, what would today be described as 'culture shock' albeit in its most benign form. For example, less than 30% of us had enjoyed the luxury of a family car, whereas in Miami virtually everyone we met had one. Of course the escape from 'blacked-out' Britain to street lights and bright shop windows had started on our arrival at Halifax, Nova Scotia.

Like hundreds of my colleagues I had traveled across the Atlantic in five days from the estuary of Scotland's River Clyde aboard a Polish liner, the *Batori*, and returned on the French ship *Louis Pasteur*-both vessels having escaped when the Germans over-ran their respective countries. But every bit as great was our delight with the food. Although the British Armed Forces were spared the worst rigors of food-rationing American meals could only be described as "something else". I had never tasted a hamburger let alone a barbecue, and I had no idea what a "weeny roast" was and had only heard of "hot dogs".

The older amongst us – the over 21s – said they missed English beer, but "Cuba Libras" (rum and coca-cola) were the height of sophistication to young palates, and just plain unlimited iced Coke was new and wonderful.

The enthusiasm for "dating" amongst the girls of our age was an unexpected pleasure. There were broken romances and a few happy marriages. I, myself, married a U.S. Army nurse whom I later met in England. Therefore, although we returned to the United States many times after the War, the priority was to spend time with her family in New England rather than visit old friends in Oklahoma, but I am still in touch with Bob Swanson, the man who taught me to fly and sent me solo in less than six hours.

Our BFTS training was exemplary. In fact, I believe it was the finest of its time. As this book explains, our flying instructors were American civilians from every aviation background, some of them scarcely older than most of us.

They wore the smart civilian uniform of the Spartan School of Aeronautics and because of that we enjoyed a different relationship with them from that of student Royal Air Force pilots trained by RAF instructors or more particularly, those wearing the uniform of the US Army Air Corps or Navy. Though full of respect and indeed, admiration, it was certainly less formal.

This book quotes an American General as saying that the dormitories of British pupil pilots were 'like pig-sties'. Whatever prompted that offensive remark it was certainly not 3 BFTS, nor I suspect any other. We were a highly disciplined unit of the Royal Air Force with a distinguished Squadron Leader responsible for everything on the ground, and a Wing Commander, with a DFC from the Battle of Britain supervising all air-borne activities. There were also two experienced non-coms (RAF pre-war regulars) with immediate responsibility for parade-ground drill and discipline. However, we were all so mad keen to get our wings and into action that the most serious offense was generally being late back from a comparatively rare night on the town. It was hard and demanding work, a test of courage and determination which should not be overlooked in this largely social record of two British Flying Training Schools.

— July 2006

The first fleet of Stearman PT17 primary trainer aircraft arrived in Ponca City from their Wichita, Kansas manufacturing plant. Built by the Boeing Aircraft Company, more than 10,000 were produced before the end of World War II. *Background photo Courtesy Ponca City Library and inset photo Courtesy Albert Sims*

CHAPTER ONE
World War II Comes to Britain

Less than a half century after great crowds of pioneers raced into the famed Cherokee Strip Outlet in Oklahoma Territory during the run of 1893, Ponca City's first and second generation residents raised their eyes and gazed north, searching the blue skies above them. They squinted, straining to catch sight of the aircraft which they could hear approaching in the far distance. As it always is in late summer, the temperature was sweltering on that August 25, 1941 day.[1] The steady humming of distant engines, gradually increasing into a cacophony of rumbling noises, intensified as many low-flying aircraft came into view. It was at that exact moment that Ponca City finally realized that a new breed of pioneers was about to set foot in their little town on the Oklahoma prairie. The British were coming! Three weeks earlier, a similar scenario had already played an opening performance at Miami, Oklahoma, when No. 3 British Flying Training School opened its doors for training of British pilots on July 13 at Spartan School of Aeronautics.

The loud droning hum of the first Boeing Stearman PT17 open-cockpit training planes surrounded Ponca City air space as they were delivered from their Wichita, Kansas factory.[2] These new, bright, yellow and blue aircraft, covered with canvas and free of radios, were flown at very low altitudes directly over Ponca City's homes and farms, signaling the fact that the much anticipated Royal Air Force cadets scheduled for training would follow quickly. The main question on everyone's mind was obvious,"Why were they traveling such a long distance to learn to fly aircraft?"

Adolph Hitler, promising the German people who had been defeated in World War I, that he would make their country strong again, came to power in Germany in 1933. Soon he was threatening neighboring countries.[3] During 1935 and 1936, the Royal Air Force, always on watch for the security of the home defenses, began a planned expansion which was designed to increase and improve units of the Home Defense while also improving service conditions for airmen and officers. The goal was to increase the officers and airmen from 45,000 to 50,000 in just one year.[4] Some Brits urged this hurried expansion because of the ever increasing build up of German fliers which were constantly expanding in numbers. Therefore, as the British began to realize that there was a very real threat to Western Europe and the Isle of Great Britain, plans were made to improve the overall strength of the Royal Air Force (RAF).

Within months of development of the Royal Air Force's initial plan, Italy declared war against Abyssinia, thus requiring that the RAF also strengthen their defenses in the Middle East. As a result, England's home defense was significantly threatened. Although the RAF handled that particular emergency, the general public never realized the part their air force had played in the situation and how depleted RAF defenses had become. England was quite vulnerable when, at a moment's notice, another emergency might occur. British government officials held their breath![5]

Acting quickly, Britain's parliament proposed an increase in taxes which would be designated for air defense and by April, 1937, the expansion scheme was gathering momentum. As the Munich Crisis unfolded in September 1938 and as Hitler's Germany tried to seize all territorial rights to Western Europe, the RAF was even more compelled to re-arm itself with more aircraft and pilots. Despite some initial appeasement by Prime Minister Neville Chamberlain, England hurriedly continued the expansion of resources vitally necessary for the inevitable war which was to come to their homeland.[6]

On September 1, 1939, Poland was attacked by Hitler's armies. That same day Operation *Panther* was initiated by No. 1 Group Bomber Command of the RAF after Chamberlain declared war. The group moved ten battle squadrons to France, where they became the direct tactical support of the British Expeditionary Force. Extremely well qualified, they were commanded by World War I veterans with considerable flying experience. By this time, both the Royal Air Force and the Luftwaffe were well equipped and considered the most effective air forces in the world.[7]

In the United Kingdom, the fighter command realized its immediate need for an effective home defense system. At that time, about 2,000 long range Luftwaffe bombers loomed on the horizon, posing an ever present threat. Britain had only 35 squadrons to help protect the shipping in coastal waters and naval ports and bases.[8]

The Blitz

May 10, 1940 marked the beginning of the blitzkrieg (lightning war) against France, Belgium, and Holland. RAF fighter squadrons flew from early morning until dark fighting the Germans. As a result, the English suffered high casualties. Of 71 Blenheim aircraft and five squadrons, 40 did not return, marking the highest loss by the RAF to that date. More and more bombers were lost and those which remained were ordered to return to England.[9]

Some officials believed that Britain's fighters should not be sent into a hopeless French campaign. Nevertheless, Prime Minister Winston Churchill sent ten squadrons of fighters to battle and the losses were high–432 Hurricanes and Spitfires were lost. Although aircrews in England were exhausted, they hurriedly re-formed.

Less than 20 miles from Dover, the Germans were well within bombing range of any point in southern

Firemen work to douse one of many London fires after an air raid. *Courtesy Imperial War Museum, London (HU1129)*

England. Stepped-up production of aircraft was pushed to the highest possible figures because of the impending battle for Britain's survival [10] and as England lost more and more aircraft and weapons at Dunkirk, the home defense stepped up to the plate. Men, women, and children armed themselves with shot guns, pitchforks, and spades as they prepared for the invading German Paratroops. Throughout the countryside, "the green fields of Kent were scattered with old cars, poles and wires strung to wreck the aircraft and gliders that were expected to land at any time with their complement of jack booted troops and guns." However, at least the citizens knew where they stood and the sentiment of almost everyone was "Let us get on with it!" [11]

After losing nearly 300 combat-tested fighter pilots in France, the RAF was not able to replace such men overnight. Pilots of bombers and reconnaissance planes were hurriedly converted to Spitfires and Hurricanes. Flying training schools in England were pumping out new pilots on a steady basis and instructors were taken into fighting service whenever they could be spared. [12]

A little girl wears one of the gas masks used for toddlers. Produced of red rubber and blue enamel, the masks were designed with a "Mickey Mouse" appearance in order to make them more acceptable for use by frightened children. *Courtesy Imperial War Museum, London (HU33355)*

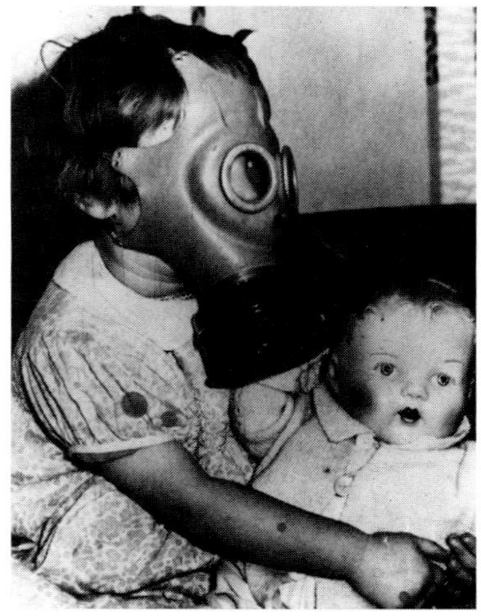

"Never in the field of human conflict was so much owed by so many to so few."

The Battle of Britain

The Battle of Britain commenced during the summer of 1940 when German attacks began on British convoys, quickly followed four days later by an all-out war campaign against the RAF. Valiant in their fight against the Germans, the Bomber command could be found retaliating each night, causing destruction to vital facilities while facing the certainty of an intensive bombing assault on the United Kingdom. [13]

Operation *Sea Lion*, Hitler's planned invasion of Britain, concentrated on airfields, aircraft factories, and radar stations in the hope that the destruction of these sites would destroy the RAF fighter command as an effective force. However, the Royal Air Force caused the Luftwaffe aircraft to suffer 227 losses in one month and 76 aircraft were lost in one day alone. [14] Utterly exhausted, the grim task of endless fighting also was taking its toll on the British.

Neighbor women chat while attending to daily chores despite the restrictions of their gas masks during a practice drill in London's South End in 1941. *Courtesy Imperial War Museum, London (HU36137)*

Nonetheless, the bravery of the Royal Air Force pilots who had fought to keep their homeland free of German invasion in early air battles was later immortalized when Winston Churchill said, "Never in the field of human conflict was so much owed by so many to so few." [15]

In an aggressive attempt to demoralize the Brits, the Germans decided to try a new tactic, shifting their attacks from the RAF to the city of London. This war targeted not only those in the military but the people themselves as Hitler's way of trying to demoralize the United Kingdom. During September 1940, Luftwaffe bombers arrived over southeast England, following the Thames River to London where they dropped an estimated 7,000 tons of incendiary devices and bombs.[16] The Woolwich Arsenal and the London docks were their first targets.[17] While the docks burned their light aided other bombers trying to site their targets. During the first night of bombing 430 citizens of London's East End lost their lives, while 16,000 were seriously injured. Blackouts were ordered for night time hours and citizens were issued gas masks to be carried "at all times." [18]

According to one of the men who trained in Miami, Oklahoma, the Luftwaffe arrived punctually every evening at ten to six and did not leave until about midnight.[19] Because of the ever present dangers involved in the driving of vehicles during the blackouts, so many automobile accidents occurred that eventually the government issued posters warning drivers to be especially careful. In East Wickham, an advertisement appeared in the paper saying: "Wait, you can see the car when the driver can't see you!" In the local Woolworth, luminous flowers were sold to be worn during blackout.[20]

During the summer of 1940 some children were sent to Ireland and Canada, the United States of America, and as far away as Australia, South Africa, and New Zealand by worried parents in order to escape the anticipated bombing. CORB, the Children's Overseas Reception Board, helped the children make the transition. With funds for travel provided by the government, and aided with donations from parents, roughly 2,500 children were moved from their homes. Nearly 14,000 others were sent privately by their own families with many going to the United States for protection.[21]

Total blackouts, required by the British government, were so dangerous to people and livestock that an Essex farmer who owned a herd of black cows painted white stripes on them so they would be visible to motorists after dark.

Courtesy Imperial War Museum, London (HU36167)

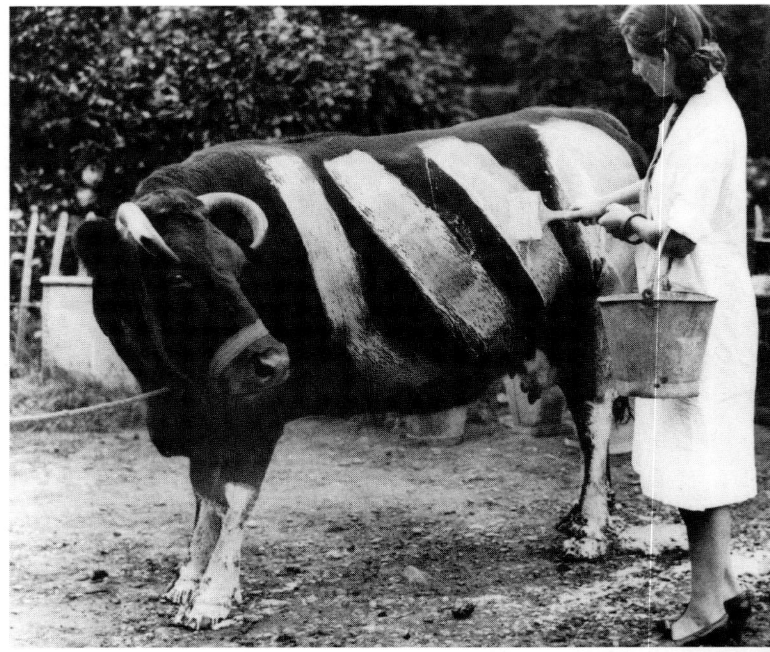

Tania Long, a war correspondent for the *New York Herald Tribune*, sent her son back to American soil for safety. In the third week of September, 1940, she reported that a ship bound for Canada, the *City of Benares*, had been torpedoed by a German submarine off the coast of Ireland. Due to heavy winds, only a few lifeboats were able to withstand the weather. About 48 hours after the ordeal began the HMS *Hurricane* rescued the survivors and returned them to Glasgow. Although several reports conflict on the actual numbers who perished, from seven to thirteen children were rescued, while 77 to 83 other children died only days after kissing their parents good-bye. Following this disaster, CORB called a halt to its operations and by October, 1940 it ceased to exist.[22 23]

The safety of the children in London and the surrounding villages was of grave concern to the British government. A few days before September 1, 1939, as Germany invaded Poland, an estimated 1.5 million British children and other citizens were moved to the countryside where they were boarded with strangers, lodging either in homes, schools, or other holding areas.[24] Most were children who took only the clothes on their back and a small bag of possessions such as tinned corn beef and margarine as they left on the trains which traveled out into the countryside. Some were randomly assigned billets while others were passed to foster parents along the way. Another two million Britons left these cities with the help of relatives and friends. During this time about fifty percent of English youth and thirty-three percent of the children of Scotland were evacuated, although only for a short time. Within a year, most had returned home because the anticipated bombings had not begun. A second round of evacuations occurred about September 1940 when bombings did begin. Some of these children drifted back home soon after, although it was not at all safe.[25] Pictured is a crowd of London school children as they prepare to leave for the West of England at Paddington Station on June 13, 1940. *Courtesy Imperial War Museum, London (HU65885)*

London Takes the Heat

For fifty-seven consecutive nights in London, more than 9,000 tons of incendiary bombs were released.[26] Although approximately 250,000 people were homeless as a result of the bombs, it was apparent that the brave fighting by the RAF had stopped any hope of an actual German occupation of Britain. By October 12, 1940 Operation *Sea Lion* was cancelled.[27]

While the Blitz on London reigned terror and destruction between September 7, 1940 and December 31, 1940, 13,339 Londoners were killed, while another 17,937 suffered serious injuries.[28] Despite their losses, British royalty did not leave London and often, King George VI and Queen Elizabeth were seen visiting with rescue parties. Women of London joined the Women's Home Defense and were instructed by members of the Home Guard on how to use a rifle. Men as old as 70 and 80 joined the Home Guard and were given the duty of searching out enemy parachutists.[29]

Bombing continued and by the middle of November other communities were attacked. They suffered severe losses as Hitler's troops dropped fiery incendiaries in the middle of cities, often on Sunday evening so that more people were at home. Liverpool, Birmingham, and Coventry were bombed seven times in one month before the Germans turned their attention back to London. During the "Baedeker raids" the ancient heritage towns of Bath, York, Canterbury, and Norwich were targeted as well as the coastal towns of Dover and Eastborne. Somehow, the people kept going about their business and for the most part, life went on as usual.[30]

Flying Bombs

On May 10, 1941 London suffered the biggest bombing raid of the Blitz with more than 700 tons of bombs and thousands of incendiaries hurled on the city.[31] Although this was the last of the major raids on London up to this time, more raids began again in June of 1944. The "Doodlebugs" had arrived. These V1 flying bombs were liquid fueled and carried a 2,000 pound warhead. The "doodlebugs" would fly until their fuel was exhausted and then crash to earth. The only warning British citizens had that they might be in peril was when the buzzing sound of the bomb ceased.

LEFT: Many families placed Anderson bomb shelters in their gardens and planted crops on their roofs because nearly every square foot of soil was used to help feed the families of London during this perilous time.[32] Their harvests included tomatoes, rhubarb, and other vegetables. *Courtesy Imperial War Museum, London (HU63827A)*

BELOW: This Anderson shelter took the full force of a bomb. In the background, people are salvaging food from a wrecked grocer's shop. *Courtesy Imperial War Museum, London (HU48554)*

Shelters

Three types of bomb shelters were used: the Anderson, the Morrison, and public shelters. Anderson shelters were most popular, with 2,300 distributed by September of 1940. Eventually, more than 2.5 million were sold.[33] A nominal fee was charged for most of them. Each shelter held up to six people and was fitted with bunks built on either inside wall. Made of corrugated steel, sheets were placed over a shallow pit measuring about four feet by ten feet, which was floored with wood. A 15-inch layer of dirt covered the shelter with 30 inches on the sides and back. Not everyone had one of these shelters because many had no garden, or yard, in which to build one.[34] Albert "Bert" Sims, Darr School of Aeronautics, Course Number 7, remembered well the large hole his father had to dig by hand to install their shelter. It was used quite often during the war and Bert's sister, Edna, four years younger than himself, shook with terror the first time they had to spend the night in it. Although she handled the situations better as they continued to be bombarded, the family never forgot the terror of the bombs overhead.[35]

The Morrison shelter, provided at no cost to the people, was used mostly indoors. With its steel roof and wire mesh sides, it was just big enough to allow one or two persons to sleep. During the day it could be used as a table. About five million were distributed by November, 1941. These were later used to protect the people from the V2 weapons.[36]

Public shelters were erected at the end of streets or under railway arches and were constructed of brick and concrete. Others used public churches, basements, and boiler rooms to protect themselves. Trench shelters were sometimes dug in public parks and gardens and could hold about 50 people each. However, they were unpopular because water tended to collect in them.[37]

A Morrison shelter was provided at no cost to citizens. Many used it as a table during the day and slept in it at night.
Courtesy Imperial War Museum, London (D2055)

The underground train stations also were used during this period. The majority of people were determined that life should go on, even when many had lost their homes and some of their family to the bombs. Still, they had to have some place to shelter. Eventually, some of the tube stations were adapted for use by the people and toilets were added. To avoid flooding due to damage from bombs, flood gates were installed and special sensors were placed in the Thames River so that nearby mines or bombs might be detected. Even government officials, including the War Cabinet, used the tubes as a meeting place in order to escape the bombing. As safe as they seemed, the tubes experienced damage by bombs and hundreds of people were killed or injured. A 1940 census conducted by the government revealed that four percent of the people were sheltering in the underground system, nine percent used public shelters, and 27 percent used domestic shelters. Many others took their chances and used their furniture as protection as they stayed in their own homes during the bombings.[39]

Many Londoners used the tube as their air-raid shelters although it was not encouraged by the Minister of Home Security, Herbert Morrison. Morrison felt that children could be hurt and that the sanitary conditions of the tube were unsuitable. He also feared that people would not come out even during daylight hours and that everyday working life would cease. Still, more than 150,000 citizens of London paid the 1 pence ticket to enter the tubes for the night.[38] Pictured here are people sleeping and living in a crowded underground train passage.
Courtesy Imperial War Museum, London (D1568)

V2 Bombs

The V2 bomb appeared for the first time on September 8, 1944 when Londoners heard a strange explosion. In the following two days about a half-dozen explosions were reported each day and people were told by officials that gas main explosions were to blame. Citizens of London suspected otherwise. In reality, what they heard was a great bang followed by a roaring, rushing sound, followed by yet another loud bang. Just before an explosion, white flashes of light could be seen blazing across the sky. The new rockets were 45 feet long, weighing about 14 tons, and they traveled faster than the speed of sound. It was reported that 518 V2 rockets exploded in London and 2,724 people died while another 6,000 were injured.[40]

Clearly, the Battle of Britain and subsequent attacks on England demonstrated to the RAF that they desperately needed training schools far from the airfields which were being heavily used for operational purposes in their homeland. It was obvious that they could not afford to use their own pilots and planes for this training because it would burden their manpower and the amount of planes they would have available for fighting the Germans.

A man takes shelter in an underground crypt at Christ Church, Spitalfields, November 6, 1940.

Courtesy Imperial War Museum, London (D1511) Photo by Bill Brandt

Training Scheme Developed

By the spring of 1941 a unique scheme for the training of RAF pilots was developed by the United States and Great Britain. President Franklin D. Roosevelt felt compelled to help the United Kingdom as he believed that the United States should serve as a "great arsenal of democracy."[41] A major debate over foreign policy arose arguing whether or not the United States should provide aid to Great Britain or whether they should remain neutral. President Roosevelt believed that America would eventually be an active participant in World War II, and on March 11, 1941 he signed a new military aid bill, the Lend-Lease Act.[42] Roosevelt's plan was to loan Britain the necessary supplies and weapons at no cost but it was to ask that Britain return the supplies or pay for them after the war. This law also allowed the United States to lend-lease supplies to other countries whose security was vital to the defense of the United States. About 58 percent of approximately $25 billion worth of war materials went to Great Britain, while the rest was distributed to Russia, France, China, and a few other countries.[43]

On August 28, 1940, Army Air Corps Colonel Carl Spaatz, after observing the Battle of Britain, suggested to the British that he would recommend the training of pilots in America, citing favorable training conditions in Texas and Oklahoma.[44] In New York, British Under-Secretary of State Captain H.H. Balfour met with United States Treasury Secretary Henry Morgenthau, who had been assigned by Roosevelt to coordinate aid to England.[45] Civilian flying schools were the subject of that meeting between the two men.

To keep the secret about the training of Britons in the United States, Spaatz laid down a series of rules for the students who would enter. No one was to enter the United States in uniform but students would arrive in the United States from Canada in civilian clothing. They would not receive any active training in live gunnery or bombing practice.[46] Because the United States anticipated that England would need its help, it was no surprise when Churchill sent letters to President Roosevelt at the close of 1940 telling of Britain's peril after their great losses from the Blitz. He reported by letter on December 9 that "The moment approaches where we shall no longer be able to pay cash for shipping and other supplies."[47] Thus, when the Lend-Lease Act was signed into law, America already had a blueprint for the future. From that time on, President Roosevelt and Prime Minister Churchill kept in constant contact, often meeting in person with one another.

General "Hap" Arnold, the commander of Army Air Forces during World War II, designed a plan for six civilian owned flying training schools within the United States which would provide "All-Through" schools, indicating the three phases of flight training: primary, intermediate, and advanced, each taught at the same location. This was a new idea at this time.[48] In 1940, prior to this scheme, the British Commonwealth Air Training Plan, at times called the Empire Air Training plan, was set into motion. Selected locations for their training helped British pilots and aircrews to be trained in some of the Commonwealth countries including, but not limited to, Canada.[49]

At these six "All Through" school locations, which were to be located in the lower United States, training progress would benefit from increased flying days due to mild weather and wide open skies needed for practice, without the danger of training cadets being shot down by the Germans.[50] Arnold recruited World War I pilots to help him implement the scheme. They included Major C.C. Mosely, Major William F. Long, H.S. Darr, Captain Max Balfour, Oliver L. Parks, and Claude Ryan. He also solicited help from E.S. Sias and Alan Hancock, top non-military flyers.[51]

Arnold explained that these schools would offer a twenty-week course with about 50 men per course. The civilian operators would provide all facilities at the school, and in return, they would receive $25.00 per hour for primary training with a rate of $35.00 per hour for advanced training instruction. The British boys would be trained by American civilian instructors with general supervision by RAF personnel, using a syllabus precisely following the RAF training program used in Britain.[52] One United States Air Force officer was assigned to each base to oversee the American cadets.

Beginning in May, 1941, under a "shaky" contract with the United States' government, American civilian contractors agreed that they would provide a large part of the flight training for the British fliers. Because the United States was not providing funding for defense, the eight men literally gambled all they owned after Arnold gave his word to ask congress for help to repay them. Several of the men even "hocked the family jewels" in an effort to gather upwards of $500,000 to invest in opening these schools. Less than six weeks after they joined forces, those six schools were open.[53]

Where would these schools be located? The six civilian schools established were: No. 1, Terrell Aviation School, Ltd., Terrell, Texas; No. 2, Polaris Flight Academy, Inc. and Mira Loma Flight Academy, War Eagle Field, Lancaster, California; No. 3, Spartan School of Aeronautics, Miami, Oklahoma; No. 4, Falcon Field, Mesa, Arizona; No. 5, Riddle-McKay Aero College, Clewiston, Florida; and No. 6, Darr School of Aeronautics, Ponca City, Oklahoma.[54]

World War II British Flying Training Schools located in the United States. *Courtesy Paula Denson*

How Royal Air Force Cadets Came to Oklahoma

The first course of fifty British cadets arrived in Ponca City on August 26, 1941.[55] It was apparent that the eager young men, with an average age of 19 or 20, welcomed the warmth and friendliness offered by the people of Ponca City. As a result, it was no surprise when it was reported that "they took to Ponca City without resistance."[56] After the first 50 arrived in Course 1, more courses were added at seven week intervals. Several of the first courses to arrive had entered the United States immediately after being processed at No. 1 Manning Depot, Coliseum, Canadian National Exhibition. This was a Royal Canadian Air Force School located at Toronto.[57] When Course 12 arrived, its roster listed 80 RAF and 20 United States Air Force cadets, who arrived at nine week intervals.[58]

Who were These Fliers?

Life had not been easy for the young men who found their way to Oklahoma. Born in the late teens and early 20s, these young men of Britain, Scotland, and Wales, already had experienced what the war with Germany had done to change the lifestyles of their families. By 1940, a blackout covered the city of London and the surrounding countryside. People used no lights after dark and all food, clothing, and furniture was strictly rationed. Women became creative in offering menus void of meat and sugar. They also gave up their extra pots and pans so that they could be melted down for the purpose of building new aircraft. Books were recycled and used to prepare shell casings. Everyone wanted to contribute their share to the war effort.

Gordon P. "Dave" Davis, a cadet trained in Miami, lost an uncle in the first few months of the war when the destroyer on which he was serving was torpedoed with only a few survivors left to tell the story. Dave wanted to avenge his uncle's death and decided to join the RAF in 1942.[59]

Many other reasons prompted young men to join. Although many of the prospective RAF cadets lived in the countryside and did not endure all the daily struggles experienced by the young men who lived in the cities, those who did had to adapt to the inconveniences of the war. During the nightly Blitz, fire bombs and fire-watching rosters following a full day's work or a restless night cramped in a garden air raid shelter also precipitated the men's actions to join the RAF. In addition, the strict food and supply rationing endured by Britain and which was brought to the citizens at the enormous cost of the lives of many Merchant Navy sailors was another primary catalyst in the response of so many young men.

Kenneth W. Ross, who trained in Miami, Oklahoma, is a good example of the type of men who volunteered to become RAF fliers. Kenneth was born in 1924 in Scotland. On a holiday morning in the spring of 1942, he took a long walk to Glasgow and volunteered for pilot training in the RAF. He decided that he must take his turn in the fight against Germany but did not tell his parents what he had done until later. Like many others, he was too young at the time and subsequently spent more time studying before taking entrance exams. Eventually he was called for duty and sent to the United States for training.[60]

There were others who wished to join the Royal Air Force. However, many of them were delayed from entrance because their civil jobs were placed on the Schedule of Reserved Occupations — ones deemed highly important to the defense of the country. Ron Wakeman, later to become a member of Course 8 at Darr School in Ponca City, was not called to the RAF until January, 1941 because he was employed by the Short Brothers Flying Boat and Seaplane Builders of Rochester, Kent, where he worked on engines, airscrews, and hydraulic systems for the RAF Sunderland Flying Boats.[61]

A group of English Air Training Corp youths, including John Cal Price (first row, far left), are shown standing at attention before civilian members of the Lewisham Borough Council on a cold Sunday morning in 1942 or 1943. They are at Picton House, Lee Terrace in Lewisham. Picton House was left derelict after air-raid damage from the Blitz which started on London in September, 1940. The Air Training Corp formed in February, 1941 but John Price was not called up for training in the RAF until March, 1943.
Courtesy John Cal Price

Youth in the British Air Training Corp are shown marching in England as their commander reviews their parade. *Courtesy Albert Sims*

Likewise, Bert Ginsberg, Darr Course 17, was not allowed to join until later because his job repairing blind flying instruments was critical to the war effort. However, at this period of the war, the shortage of pilots had become so acute that the schedule was amended to allow an intake of men to be trained as pilots and navigators.[62]

To prepare for their entrance to the country which was to host them, young men from England and other commonwealth countries, still pink behind the ears, were issued a small blue book titled *Notes for Your Guidance*.[63] These were probably issued at the main dispatch station at Heaton Park in England. What the little book contained was information intended to guide the young men in their reactions to various questions and predicaments which might present themselves during the men's stay on the American prairie. A high standard was expected from the men on many levels including politeness, an open-mindedness to new situations, and a willingness to learn new ideas from a culture different than their own. They were encouraged to listen more than to speak. The young cadets also were urged to learn more about the history of the United States so that they could understand its economic and political systems and perceive the vastness of the host land. Language, both English and "American," was emphasized as being different in many ways and so the lads were cautioned to carefully consider how they phrased their subjects,

A typical scene shows cadets at the Ponca City train station. The station was always busy as 17 courses arrived and departed during the school's existence. Pictured left to right are sergeants: Hugh Fielding-Johnson, Roland Dashwood, Mike Flood, Owen Lloyd Davies, and an unknown cadet. *Courtesy Dr. and Mrs. J.E. Fuller*

The *Queen Elizabeth* was one of the British luxury liners converted for war duty. It is surrounded by tenders and small craft used to protect her as she arrives from America on November 23, 1944. The great ship was typically loaded with up to 5,000 to 6,000 soldiers as she transported them from England to the United States and back again. *Courtesy Imperial War Museum, London (A17862)*

less the men provide more humor than was necessary in some instances and saving themselves some future embarrassment. Sports was a topic given quite a lot of attention in the booklet as it explained that American codes of sportsmanship were different from the English ways and that they should respect these codes, even if they did not always understand them.[64]

On the other side of the pond, American servicemen entering Britain in 1942 were issued similar instructions. The history of Great Britain was reviewed and soldiers were reminded that Britain had been in war since 1939 so there should be great understanding as to why there were no "frills" in many areas of daily life. The customs and manners of the people were explained and soldiers were instructed about driving on the left side of the road and about drinking warm beer. Sports like soccer, cricket, and "rugger" were explained to them, just as the little blue book issued to the Britons explained American football. In short, the Americans were reminded that they were "coming from a country where homes were still safe, food was still plentiful, and lights were still burning."[65]

Soon after Ponca City was chosen to be one of the British Flying Training Schools, Mrs. Eleanor Roosevelt contacted Mrs. Gladys Dutton by letter, asking a favor. Mrs. Dutton shared the request from Eleanor Roosevelt at a meeting of about 300 various church and civic club women in Ponca City. Mrs. Roosevelt stressed to the women the importance of breaking down the barriers the men had due to preconceived notions that Americans were different from the English. She told the group that they could do more, as mothers, to ensure lifelong friendships and tell the world how the Americans really were by inviting the young British men into their homes and hearts. Mrs. Roosevelt wanted the men to see a clear picture of what Americans were really like and that all Americans did not live the lives of glamour portrayed on the Hollywood movie screens. Mrs. Roosevelt explained that these men already had suffered atrocities to their homes and families and that because of the strict rationing, some of their diets had been lacking of anything tasty and nutritious since rationing began in 1939. She wanted the boys to feel like they were at home with their hosts and if that meant that the women should let the men scramble eggs and make toast and wash dishes, then so be it![66]

During the 40s, the "Blue Laws" were enforced in many American states. Businesses were not open on Sunday, one of the free days which would be enjoyed by the young men.[67] This was the perfect opportunity for the citizens to extend their invitations of hospitality to the lads and a way in which to fulfill Mrs. Roosevelt's requests. Ponca City's women took her plea to heart and opened their homes and their hearts as they prepared for the arrival of the first of seventeen courses to come to Ponca City.

Three unidentified cadets enjoy a sled ride at Moncton, New Brunswick before their departure to sunny Ponca City. *Courtesy John Cal Price*

The men of the RAF came from an eclectic variety of home environments as they joined the Royal Air Force and commenced their journey to North America. The young cadets were boarded on ships in the dark of night and traveled to unknown destinations across the North Atlantic. They were contained in military convoys, flanked by destroyers, because German submarines and their deadly torpedoes were an ever present threat to their lives. Excitedly, the men made guesses about their final destinations. Most of the chaps guessed that they were being sent to Canada and they were right. However, many of the ships arrived first in New York City, New York or Boston, Massachusetts, before their occupants were put on rail cars and transported to Moncton, New Brunswick. Dennis Stone, who was later assigned to Course 17 in Ponca City, was so thrilled by the approaching sky line of New York City as the *Queen Mary* neared the harbor in December, 1943 that more than 63 years later, the memory of a bright blue sky with a silhouette of silvery needles very gradually growing larger as it rose over the horizon was still very crisp. As each hour passed these silhouettes enlarged until Dennis eventually realized that the largest city in the United States was coming into view.

For those RAF cadets who arrived in New York City and had time to explore before they left for Moncton, a brochure was distributed which listed information for the RAF transient personnel. It included addresses of various sports clubs, WAAF Clubs, places to stay, and department stores such as John Wanamaker's, Macy's, Bloomingdales, Stern's, Lord & Taylor's, Bonwit Teller's, and Saks' Fifth Avenue. A variety of coupons were offered and many laundries, banks, hospitals, and camp facilities at Fort Hamilton also were described in the information.[68]

Another cadet, S. J. Eaton, made his trip across the Atlantic Ocean to the United States and later to Moncton via the *Highland Brigade*, a flat bottomed ship that had been used to carry meat on the River Plate. The cadets had to hang their hammocks from the meat hooks in the ship's hold during their crossing. They were flanked by two destroyers as they zigzagged their way across the Atlantic Ocean.[69]

To celebrate his 21st birthday while training in Oklahoma, Mrs. Parker baked a birthday cake for John Cal Price, center. He is flanked by Gordon Huntley, left, and Dennis Stone.
Courtesy John Cal Price

Albert "Bert" Sims told of a similar story of his crossing of the dangerous Atlantic on the S.S. *Banfora.*[70] Billeted near the stern of the ship and on the lowest level, Bert slept in a hammock hung from meat hooks, swinging from side to side with the motion of the ship. Many of the other men who shared that portion of the ship slept directly on tables, laying their mattresses on top of them. To pass the time, the men played cards. Several days before their arrival in Halifax, Nova Scotia, they took turns showering in cold, salty sea water.[71]

Moncton, New Brunswick, the location of the first "home" assigned to the cadets arriving on the northern continent,[72] was a dispatching station. Because authorities on the base were apparently under the misconception that the cadets would be sent to areas where temperatures were frigid, they issued the men grey flannel Montague Burton dress suits, a long sleeve shirt, woolen socks, and underwear. It was recommended that the men purchase a heavy overcoat with their clothing allowance. Taking the officials to heart, most of the men followed the advice.

Those who were assigned to Oklahoma did not know where the towns of Ponca City and Miami were located but when they were told they were going on a train ride across a wide expanse of the United States of America they could hardly believe their good fortune. Some would travel to Number 3 British Flying Training School in Miami, Oklahoma while others were assigned to Number 6 British Flying Training School in Ponca City, Oklahoma.

Three Ponca City British cadets visit the Oklahoma State Capitol Building, wearing their grey flannel Montague Burton dress suits. Coming to Oklahoma and walking in the heat made the men quite uncomfortable. After the United States entered World War II, the men were allowed to dress in their uniforms. *Courtesy Mariann Bryant*

The Atchison, Topeka, and Santa Fe Railroad carried thousands of men to their destinations across the United States, including Ponca City and Miami, Oklahoma. John Price at the side of the engine which brought him to Ponca City. *Courtesy John Cal Price*

CHAPTER TWO

Number 6 British Flying Training School Darr School of Aeronautics Ponca City, Oklahoma

In Oklahoma, Legionnaires at the Eighth District Convention held in Enid favored the safe delivery of all-out aid to England and advocated that defense labor be put on a full-time basis.[73] They were not alone as communities all over the United States voiced the same opinion. Responding quickly, one of many business men who entered a bid for one of the flying training schools was Harold S. "Hal" Darr.

Initially, Darr was a Lieutenant in World War I and an instructor for the United States Army Signal Corp cadets, teaching in Texas and Florida. He based his flying businesses in Chicago, Illinois and at the time the British Flying Training Schools were created, Hal Darr was the chairman of the board of Braniff Airlines.

After World War I, Darr plied his skills in barnstorming, a practice popular in that time. Later he established the first Darr School of Aeronautics, located at Glenview, Illinois, where he became a dealer/distributor for Stinson monoplanes. His old friend, Eddie "Early Bird" Stinson, had trained in World War I under another flying buddy of Harold Darr, Lieutenant Clarence F. Cornish. By 1927, Stinson was a successful producer of aircraft. As a founder of the Aeronautical Training Society, he visited Washington, D.C. and helped formulate the plans for the future mass aviator training project.

In the winter of 1940 and 1941, Hal Darr helped establish an Army training contract base at his Glenview school where Stearman PT17 trainers and flying cadets were sent to Darr for primary training.

Realizing the need for better flying conditions, Darr established the Darr-Lodwick primary training operation at Lakeland, Florida. His right hand men included Paul F. Victor, Henry "Hank" Jerger, and Berney DeWitt. These three instructors had been chosen to teach flight instructors the new standardization of techniques required in future flight training programs for both military and domestic pilots, as required by the Civil Aeronautics Authority. This was later known as the Federal Aviation Administration. Jerger, Victor, and DeWitt worked to run one of the first primary contract fields at Curtiss-Wright, Glenview, Illinois for Hal Darr.

After Curtiss-Wright closed, the three instructors were sent to Albany, Georgia where a new primary school at Turner Field was established. By August 1941, Augusta, Georgia was the sight of Bush Field, a new primary school set up for Hal Darr. The instructors then proceeded to Kelly Field where they were taught advanced training so that they could again be assigned to Hal Darr for the new British pilot training program.

Darr sold his half ownership to Al Lodwick in July 1941, took the capital and began the establishment of a new base at the Ponca City, Oklahoma Municipal Airport.[74] Enid, Oklahoma was fiercely competing for the school but the Ponca City Chamber of Commerce was negotiating with city officials and with a private company. They would establish the school in Ponca City under a government contract.[75]

Earlier in the summer, Ponca City had officially been approved for the British pilot training school to be run by the Spartan School of Aeronautics. Because a government ruling stated that training schools could not be located at commercial airports and a change in the mood of Captain Maxwell W. Balfour, head of Spartan, the school's location was changed to Miami, Oklahoma.[76]

Harold "Hal" Darr, third from left, arrives with various RAF officials as they make an inspection of the new British Flying Training School established in Ponca City in August, 1941. *Courtesy Friends of Marland's Grand Home Foundation*

Darr School of Aeronautics as it looked soon after initial construction on the new campus. *Courtesy Bret Carter*

Participants of the Ponca City Chamber of Commerce aviation committee who were bidding for a school were Col. T.D. Harris, a Conoco official, Lee Haugen, Dr. C.E. Northcutt, L.H. Wentz, and J.K. Cleary.[77] Mayor Frank Overstreet and Commissioner Guy Conner also played a role in the negotiations while trying to win the school.[78] Merchants were absolutely eager to have the school locate in Ponca City because it would boost their economy tremendously. The chamber reported that Darr School would pay $100 per month rental to the city and would pay the regular industrial rate for the water and electricity which it used. The existing flight services, including Braniff Airways, Ponca City School of Aeronautics, and the Continental Oil Company would continue to use the airport.[79]

Ponca City offered the north end of the Municipal Airport after increasing the size of the airport from 240 to 520 acres.[80] A few days later they leased 484 more acres for the main flying field at the Ponca City airport. They increased the length of the runway to 1¼ miles. Eventually there were two paved runways, each 3,500 feet long. Forty acres was leased from O.A. Edwards, 80 acres from W. Waddelow, and 180 acres from Eugene Wetzel. Later, an auxiliary field, located four miles west of the northwest corner of the Continental Oil Company tank farm was added, including 160 acres leased from Mrs. Catherine Pfeiffer of Orlando, Oklahoma.[81]

When Darr signed the contract with Ponca City on August 5, 1941 for a training enrollment of two hundred, he already had three other military flight training schools in operation. However, this was to be the first school built exclusively to train British pilots. Construction began on the school on August 7, 1941, less than twenty-one days after plans were completed.

Flying fields with large hangars were planned not far from the campus but Darr knew that other auxiliary flying fields would have to be found several miles beyond the schools. Typically, they were large grass fields. In primary training, half the planes would be flown to the auxiliary fields each morning while the other half of the men would ride there in buses. In the afternoon, the reverse would take place.[83]

Construction of Barracks Begins

Dick Sherbon, a local contractor, was hired to provide construction of the new barracks. Designed by Joe Cannon, a Ponca City architect who had previously designed the Ponca City Library, Ponca City's mausoleum, Roosevelt School, and St. Mary's School, the barracks were designed in a semi-southern style featuring white walls, green slate roofs, and small porches with white colonnades.[84] Each barracks was

TOP: Initially, Stearman PT17 aircrafts, manufactured by Boeing Aircraft, were used for primary training at both Oklahoma British flying training schools. *Courtesy Jack Youens*

ABOVE LEFT: Veteran broadcaster Raymond Baxter, OBE, trained as a cadet in Miami, Oklahoma as part of No. 3 British Flying Training School in World War II. He served as a Flight Lieutenant with 602 RAF Squadron from Glasgow, Scotland. He flew a Mark XVI Spitfire as he participated in the March 18, 1945 precision low level attack on the V1/V2 headquarters in The Hague. The code name for the attack was "Big Ben" because British intelligence discovered most of these incendiary rockets were aimed at central London. *Courtesy Raymond Baxter*

ABOVE RIGHT: Pilot trainees graduated directly from the Stearman PT17 to the Harvard AT6 for their more advanced training after the Vultee BT13A was eliminated as a training aircraft. Many of the student pilots considered the Harvard AT6 to be one of the most "fun" airplanes flown. *Courtesy Dr. and Mrs. J.E. Fuller* designed with two large dormitories. A recreation room split the center of each barracks and offered a bathroom wing at the back. The recreation room was equipped with a stage, phonograph and radio, ping pong table, and a piano.[85] Double-tiered bunks furnished the rooms with large lockers dividing each room.[86]

Construction rolled into high gear in early August and Mr. King announced that the Ed Cooley Plumbing Company and the Braker Electric Company, both local firms, had won bids in Ponca City.[87] A delay in construction seemed to be a problem for a short time. On August 11, Pat Schlesinger, head of the city water and light department, reported that the city could not get the four inch pipe needed for the school.[88] They were given a priority rating of E, which was assigned by the office of production management in the government.[89] However, by August 20, 4,000 feet of water pipe was scheduled for shipment from a Birmingham, Alabama steel mill after an appeal was made on behalf of Ponca City by Congressman Ross Rizley.[90] By this time, the Northern Oklahoma Gas Company had started laying gas lines and electrical connections also were being installed by the city light department. Sewer pipe was being laid at the same time.[91]

By August 25 it was announced that the first six Boeing Stearman PT17 primary training planes for the Darr School of Aeronautics would arrive from a Wichita, Kansas factory and that the first course of RAF aviation cadets would appear within a week.[92] By August 29, the total number of training planes had

A group of Course 16 pilots prepare for take off in a Harvard AT6. Pictured left to right are Norman Griffiths, unknown cadet, Joe Moyle, and Grant Callister (USAAF). *Courtesy Phyllis Mann Wiles*

This former Conoco triangle was a well known logo throughout the United States before it was replaced by a newer design. *Courtesy Paula Denson*

Continental Oil Company kept the aircraft used by Darr School of Aeronautics fueled throughout the time the school existed in Ponca City. *Courtesy Ponca City Library*

increased to 19 [93] and by September 14, the number had been increased to 28. [94] Belonging to the United States government, the lower wings of the primary training planes featured the painted inscription, United States Army. The planes, part of the Lend-Lease program, were turned over to the Darr school for use by the British under terms of the program. [95]

Vultee BT13A airplanes, known among RAF students as "The Vibrator," [96] were brought into the training scheme for use as soon as the students completed their primary training. Produced by the Vultee Aircraft Corporation of Englewood, California, the low-winged planes were powered by 450-horsepower motors with flight speeds in excess of 200 miles per hour. These planes were equipped with radios. [97] Later, the advanced training courses saw the use of the North American Harvard AT6 aircraft produced by the North American Aviation Corporation of Englewood, California. This low-winged plane was larger and more powerful than the Vultee. Its equipment contained more flying devices, such as two-way radio equipment and all blind-flight instruments. [98]

Aviation fuel for the training aircraft was provided by Conoco Oil Company. At the beginning of 1941, and for one year's time following, the oil companies in the United States drilled 32,140 oil wells in an effort to

Shortly after the schools were initiated, the Vultee BT13A was eliminated at the intermediate level of training. Pictured are four of these aircraft flying in formation. *Courtesy Air Vice-Marshal H. Gill, CB, OBE, RAF retired*

produce the 500,000 barrels needed daily to take care of the war.[99] In Ponca City, Continental Oil Company manufactured its first barrel of 100-octane aviation gasoline, a high-performance, high-altitude fuel in 1944. Beginning in 1942, the company developed a new plant which operated as a separate refinery in Ponca City and which produced the finished aviation fuel required by the government.[100] On its first day of production it yielded in excess of 25,000 barrels for the armed forces.[101] Just days before the closing of Darr School in Ponca City, Continental Oil Company received praise from W.B. Young, Rear Admiral and Chief of the Bureau of Supplies and Accounts for the United States Navy. In record time the company had produced 25,000 drums of aircraft oil after an emergency order by the navy. It took 180 box cars to ship the cargo during that emergency and the company continued to set high standards for war-time production in the following years.[102]

How No. 6 British Flying Training School Was Run

American pilots Henry "Hank" Jerger, B.G. Dewitt, and Paul F. Victor were responsible for setting up the "All-Through" school, which offered flying day and night, seven days a week.[103] Initially, the school would start with fifty trainees and continue adding groups until the total reached 200. Courses typically lasted between twenty to twenty-four weeks with approximately 50 students per course. Each course was trained in three stages including Primary, Basic, and Advanced.[104]

A group of about 150 instructors and mechanics were expected about the same time that it was announced that the first hangar, 206 by 124 feet, would be constructed. The school would contain four barracks, a mess hall, an administration building, and a classroom building with P.W. King, engineer, in charge of construction. King previously had supervised the

A bird's eye view of the Continental Oil Refinery tank farm with the Arkansas River in the background was taken by a British cadet during his flying training. *Courtesy Mariann Bryant*

Squadron Leader H.F. Suren arrived in Ponca City with a frosty attitude. Several weeks later he felt so much more comfortable with his American associates that he relaxed and began to use their slang. *Courtesy Friends of Marland's Grand Home Foundation*

construction of three other Darr schools in Georgia and Florida. Additional land would be leased for the purpose of additional practice landing fields and would be contained within a 15-mile radius of the main airport. Instructors and trainees would take off from the main airport and fly to the practice areas for landing practice. [105]

Shortly after the arrival of the American staff, RAF Squadron Leader H.F. Suren arrived in Ponca City to accept his duties at Darr School of Aeronautics. A ten-year veteran of the RAF, he was to have charge of discipline and other matters dealing with the activities of the flight students. [106] Suren began his tour of duty in Ponca City with a frosty attitude, never speaking to anyone to whom he had not been introduced. However, after about eight weeks in the friendly Oklahoma town, he was reported to have been "slanging the language" and speaking to everyone first while tipping his hat. He admitted that he had a preconceived notion of Oklahoma and thoroughly expected to find cowboys and Indians. He was especially surprised to find the Native Americans shopping in the drug stores and congregating on the streets corners of town–and not so much like the ones he saw in the movies.

Suren, a former cavalryman, made the observation after riding in Oklahoma that "an English saddle is a bit of leather with a hoss under it and a western saddle is a lot of leather with a bit of a hoss under it." [107]

DRoss Ellis, an army air corps reserve officer and a native of Stillwater, Oklahoma, was assigned as flight director of the school. He was to supervise more than 100 mechanics and maintenance men as well as an anticipated number of 45 to 50 flying and ground school instructors. Only the Royal Air Force trainees and their own officers would live at the school barracks. Ellis, along with an estimated 150 families moving to Ponca City, would live in civilian housing after they arrived. Mr. Ellis had been a flying officer in charge of training British and United States pilots at the Darr Aero-Tech School at Albany, Georgia. [108]

Because there was a shortage of qualified air instructors, Darr School of Aeronautics, under the supervision of DRoss Ellis, announced that job applications would be taken from persons who had flying experience but who did not have an instructor rating. Up to one year prior to the United States' entry into World War II on December 7, 1941, hundreds of

While the English were in Ponca City, many took up horse back riding during their leisure hours. They were enthralled by visions of the "Wild West" they had seen in the movies. *Courtesy Paula Denson*

Army primary training bases had been established in an effort to train qualified instructors. Even they could not keep up with the demand. Ellis said, "We will give them a course and prepare them to pass the regular civil aeronautics administration exams." [109]

V.L. "Casey" Jones, previously the director at the Aero-Tech school in Augusta, Georgia was assigned the task of organizing the ground school training in Ponca City. In preparation for this task, he visited the Spartan School of Aeronautics in Miami, Oklahoma which had been opened approximately one month before. [110]

In short time, A.E. Runnels, Jr., a Ponca City high school graduate of 1938, was hired as the flight dispatcher for the school. About the same time, L.H. Baillie arrived from Hackensack, New Jersey. He would manage the business office of the school, which was initially located in the First National Bank building. [111] Francis Coleman, a graduate of the National Youth Administration Radio School at Tonkawa, was hired as a ground school radio instructor for the school. [112] By August 29, 1941 seven flight instructors arrived from the Darr Aero-Tech School in Albany, Georgia. They were John Kometer, Paul W. Anderson, R.P. McComb, E.J. Miron, M.H. Meyers, C.W. Fellers, and L.M. Lacey. [113]

At Ponca City, the American flag, facing north in front of the school, waved along side the Union Jack from a halyard atop a mast. One was raised first one

This pen and ink drawing of downtown Ponca City was drawn by Reg Kemp, a cadet at Darr School. Cuzalina's Drug Store, a favorite gathering place, is featured on the left. *Courtesy Ponca City Library*

With the Ponca City grain elevator directly behind them, four British cadets pose for their picture on one of their jaunts to town. *Courtesy Phyllis Mann Wiles*

day, with the other raised first the following day. Flag etiquette required that the American flag should fly at the right as a person faced the entrance to the school. Later the Union Jack was replaced by the RAF ensign. Designed with a pale blue field containing a miniature Union Jack in the left corner, it had red, white and blue roundels located in the right corner. British cadets were required to halt and salute both flags each time they passed them. [114] This practice also was observed in Miami, Oklahoma.

The First Course Arrives in Ponca City

As the young men in the first course of fifty arrived at Ponca City on the Atchison, Topeka and Santa Fe train, they were weary from the long ride. They had left Moncton, boarding a Canadian Pacific train bound for Chicago. There they changed to the ATSF train line. During their four-day journey, the wide-eyed young men realized what an enormous country they were visiting. When they arrived at their destination in Oklahoma, they immediately noticed that the train appeared to stop in the middle of a wide, flat, brick lined street. Looking east, they saw six or seven blocks of businesses set on either side. The train continued to roll for another two blocks and came to rest at the station, which also looked considerably different from those in England. Back at home, the men were accustomed to train platforms which were elevated.

On the evening in which the first group of men arrived, a few people from Ponca City met the Santa Fe train although the weather was rainy. [115] They could hardly wait to welcome the men to their new home in America. Two blocks south of the Grand Avenue crossing stood a three to four-story grain elevator, casting a large shadow upon the surrounding buildings, including the Ponca City train depot. Sometime after America entered the war, a large, black "V" for Victory was painted high on the side of the huge concrete tower. When Bert Sims and his companions saw the inspiring symbol they immediately felt at home because they knew that America was supporting them in every way. [116] In this manner the British chaps who had come to America to learn to fly for the critical defense of their homeland's freedom were introduced to the friendly people of Ponca City.

Peter Greaves, Course 10, related:

The day we arrived in Ponca the news must have got out for there were many to greet us at the station but the big welcome came when after being confined to camp for about six weeks, we were let loose on the local inhabitants for a whole Saturday, and what a welcome that was too – never to be forgotten or equaled. As we got to the camp gates we found an unending line of cars, to left and right, full of families waiting to take us around and show us the town. And so it was every week. [117]

Rick Brown arrived in late 1943 and Ponca City citizens were still at the trains to send off the last course and greet the new course with as much enthusiasm as they had exhibited for the first courses. He reported that "people were wonderful 100 percent of the time." [118]

The first contingent of trainees was accompanied by Flight Lieutenant T.G.M. Davidson of the Royal Canadian Air Force. Major H.F. Suren also greeted the young men who had traveled from Moncton, New Brunswick, then from Toronto, Canada, with brief stops at Chicago, Illinois, and Kansas City, Missouri. As soon as they stepped off the train, the young men were whisked away to several big yellow school buses and were said to have been grinning ear to ear. The happy young men learned that Ponca City was in the midst of their Cherokee Strip Celebration. Immediately after they were seated for their journey to the temporary barracks, the group broke into song. Several of the young men even produced harmonicas while others sang:

"I've got sixpence, jolly, jolly sixpence,
I've got sixpence to last me all my life.
I've got two pence (tuppence) to lend,
Two pence to spend, and two pence to send home to my wife . . .
No cares have I to grieve me
No pretty little Miss to deceive me
As I go rolling home."

Refrain:
"Rolling home, rolling home
By the light of the silvery moon,
Happy is the day when the airman gets his pay,
Rolling, rolling, rolling home." [119]

A Ponca City woman was quickly heard to respond in astonishment, "Why, they are singing in American!" [120]

The young men, many of whom were less than twenty, upon arrival at their temporary barracks, began to answer many questions. They quickly explained that the shiny RAF emblems proudly pinned to their hot and scratchy wool suits were made of brass, a commodity in short supply due to a shortage of all types of metals. Most metal had been consumed in the war effort. The men mentioned their short visit in Chicago where they noticed the abundance of tall buildings. Then they asked how tall the buildings were in Ponca City. Six stories stirred their interest enough for them to comment:

"There was a five-story building at home but when the 'Jerries' finished it was not nearly that tall." [121]

Two flags fly proudly atop the mast at Spartan School of Aeronautics in Miami, Oklahoma.
Courtesy Deryk Maddox

Because the barracks in Ponca City were still under construction as the first course of men arrived, arrangements were made with the Ponca Military Academy to provide temporary housing and meals for the cadets. Use of the mess hall and classrooms also was provided.[122] The school, itself, was only in its second year of operation when the British arrived in Ponca City. Major William Cox, director of PMA, together with his wife, made each new student feel quite welcome.[123] As cadets waited for training to begin, they learned to enjoy the game of ping pong at Ponca Military Academy.[124] During their stay at PMA, two young lasses were so determined to meet them soon after the first course arrived that they followed them home that evening and called them from their beds to stand at the windows where they answered questions and chatted for about fifteen minutes before they were discovered and asked to leave.[125]

While the young men in Ponca City were billeted at Ponca Military Academy they were bussed back and forth to the airport. A tradition was initiated as men of the first course completed their first solo flights. They were unceremoniously thrown in the pond near the Ponca City Hospital in celebration of the momentous flights. Some of the later courses continued to celebrate with the same tradition.[126]

As soon as possible after their arrival at the Ponca Military Academy, the new cadets managed to answer many questions put forth by inquisitive Ponca City residents who had followed them there. In the friendliness of the Conoco golf club house where they were offered cold drinks, they questioned their hosts with such questions as: "Are there many girls here?" After an affirmative, they continued, "And are there many fellows here?" After another affirmative answer, they quickly replied, "Well, I suppose a jolly bit of competition would be all right!"[127]

Although the cadets spent most of their time flying and studying, it did not take long for them to realize that almost every woman in Ponca City was a wonderful cook. To enjoy the taste of delicious home cooking was a special treat which no cadet hesitated to accept.[128] Invitations into homes, schools, and churches were constantly offered.

The energetic young men enjoyed thick T-bone steaks, creamed potatoes, lots of ice, hamburgers, all

Ponca Military Academy, a junior military school, was established in 1940 by Captain Gerald J. Cox and Captain William V. Cox, both formerly involved with the Missouri Military Academy at Mexico, Missouri. The school taught boys from first through ninth grade. The buildings were originally constructed by E.W. Marland and designed for use by his single male employees. After he lost his fortune, the property remained in the holdings of Continental Oil Company. *Courtesy Ponca City Library*

The Dutton family "adopted" many young men as they came to Ponca City. They were always well fed and were often taken on wiener roasts such as this one. Pictured with Mr. and Mrs. James W. Dutton are several cadets, including Mike Igglesden, shown on the far right. *Photo courtesy Sarah Barrington Westerheide*

fruits, apple pies, cabbage slaw, mounds of butter, and plenty of eggs.[129] At home, only one egg was allowed per month due to the strict rationing.[130] A few days after some of the men arrived in North America, they ventured forth for their first visit to a local restaurant, where several of the guys ordered "Eggs and Chips." When asked if they wanted "sunny side up or over easy" they held a conference and decided to shoot for "sunny side up"–whatever that meant. In just a short while, the men were amazed to find fried eggs accompanied by potato chips and a bottle of tomato ketchup. One of the men caught the attention of the waitress, politely informing her that she had accidentally given two eggs which they had not ordered. In response, the waitress explained that two eggs were always served unless you specified only one. From then on, the men never pointed out the number of eggs being served to them, even though it seemed extravagant.[131]

Bananas were another delicious commodity missing in England. Storefronts would display large signs announcing "Yes, we have no bananas!"[132] S.J. "Jerry" Eaton, of Course 6 at Darr School, loved the food he encountered during his stay in Ponca City. Shortly after his arrival, Jerry, along with three friends, went out for a meal. Each ordered a T-bone steak. When the first steak arrived the men asked for a knife to cut it into four pieces before they were told that the steak was for one person and that the other three would soon be served.[133]

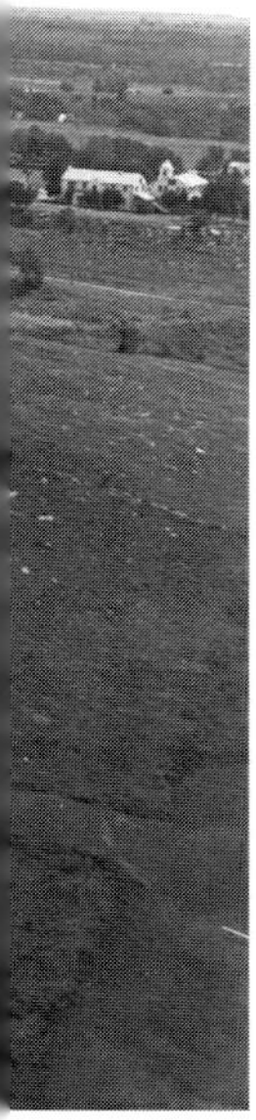

Two Ponca City British cadets treat themselves to bananas at a local grocery store. Behind them a bunch of bananas hang ready for purchase, along with a supply of Christmas trees. Apparently the weather was not cold enough to freeze the fruit.
Courtesy Ponca City Library

This cartoon illustrated the scarcity of eggs in England and was featured in the *Open Post Magazine* in Miami, Oklahoma. The official magazine for No. 3 BFTS echoed the same warm feelings that the cadets in Ponca City felt. *Courtesy Ponca City Library*

The head cook at Darr School was Charles Lambert. He was joined by cooks "Swede" Allard and Mr. Peters. [134] Another one of the men who worked in the kitchen at Darr School was Dwight John Bottrell, Jr. As a teenager, he would walk to the school each day to help prepare the food. [135] Earlier, "Swede" had been a cook for Lew Wentz, the Jens Marie Hotel, the Moose Ballroom, and the Arcade Hotel. It was not long before the student pilots asked him to place hamburgers with pickle, mustard, and onions on the menu.[136] The novice pilots also enjoyed the plentiful quantities of butter, ice, and fruits served by Mr. Lambert and his staff, who also supplied the boys with baked beans or creamed kidneys for breakfast. [137]

Squadron Leader Suren quickly discovered that the boys' fondness for sweets and rich foods was causing many stomach disorders. He cut the menu down to wholesome, plain foods. Immediately the sick list dropped to almost nothing. Suren did request of the cooks that the English breakfast specialty of baked beans on toast be included on the menu every Sunday. Lambert obliged the squadron leader's requests while he also prepared many new foods for the boys. He served chili on the menu but he did not prepare afternoon tea.[138]

Studies began almost immediately and there was little time for the men to familiarize themselves with their new hometown for several weeks. Most of the men had little flying experience before their arrival. Even more of them had no knowledge of how to drive a car. Nonetheless, they took to the air and did not seem to be afraid to learn to fly. [139] Although school policy stated that they were not to operate motor vehicles while they were in Oklahoma, many of the cadets found that the citizens of Ponca City trusted them enough to lend their automobiles.

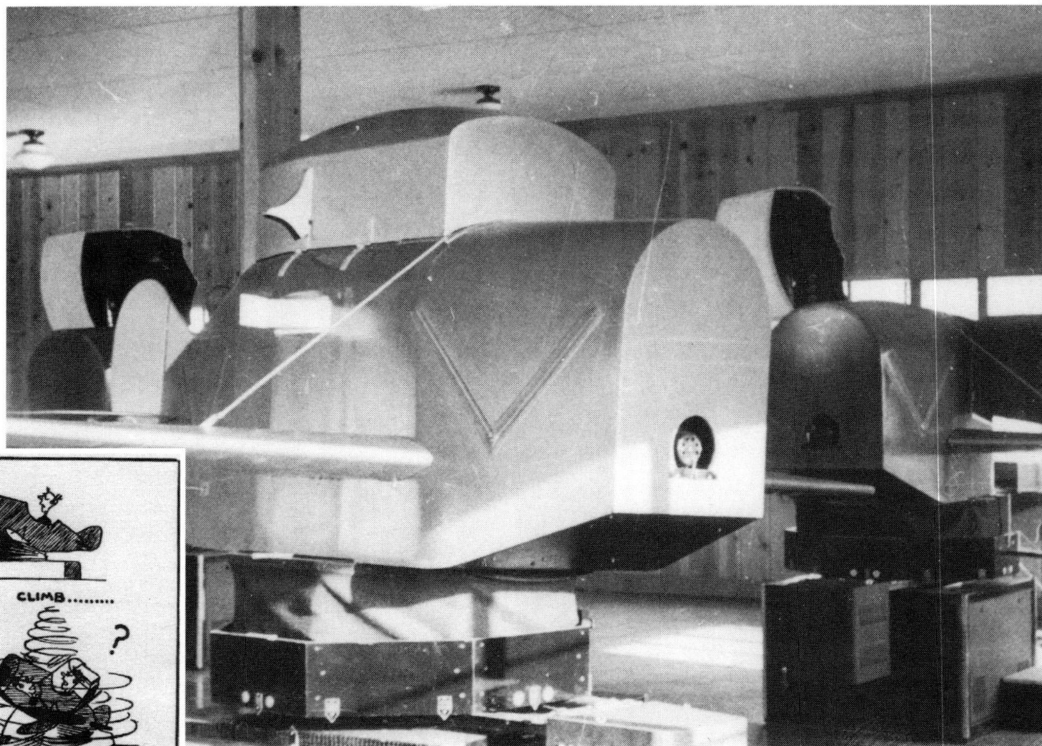

Claustrophobic cadets were never very thrilled to climb into the Link Trainers. The interiors were cramped and hot. The instructors would issue a series of flying directions and maneuvers through an intercom. Several cadet cartoonists took the opportunity to draw comical views of the apparatus which were published in the *Open Post Magazine* at Miami, Oklahoma. *Courtesy Ponca City Library*

The Important Roles of Link Trainers, Instructors, Mechanics and Women

Link training was an important part of the curriculum at all British Flying Training Schools. The Link Trainer was originally designed by Edwin Albert Link in 1929 [140] and quickly became the forerunner of the present day flight simulator. [141] It was run by a sophisticated collage of pneumatic vacuum bags and other pulleys, belts, and hoses and produced the feeling of floating on air. [142] The trainer faithfully produced movements of an airplane operated by similar controls. It did not, however, simulate steep climbs, dives, or loops. The students were able to experience "blind" flying without the hazard of being injured during training. [143]

As they sat in one of two cockpits almost identical to those found in their training planes, students found they did not always enjoy the experience, although in the heat of summer it was a pleasure to visit the Link Trainer room. A primitive air conditioning system was installed there in order to keep dust out of the delicate mechanisms of the trainers. Although the air was cooler in the building, some of the students seemed to think that the instructors took delight in putting them in odd positions after the hood was closed. [144] Actually, the instructor's table contained a tracing machine and all moves made by the student in the cockpit were recorded. A detailed plan of flight was provided by the instructor and if the student happened to err, his mistake was easily identified by the instructor. [145]

One of the Link instructors who worked at Darr School was Lillian Bowles. Another was Juliette McNiece, who later worked for the OSS in London before she married one of the American instructors,

Link instructors at Darr School pose for a group photo. Lillian Bowles Taylor is pictured on back row, second from right.
Courtesy Ponca City Library

Paul Victor. After high school graduation Lillian attended business school and immediately upon completion went to an interview for a secretarial job at Continental Oil Company. However, although she knew the typewriter keyboard intimately well, during the job interview she was seated next to another young woman applying for the same position. She, too, was quite proficient at typing. Soon after the typing test began, Lillian felt that she was doing quite well. Unfortunately, when she looked down at her paper, she discovered she had started on the wrong line of keys. Without a word but with a rather red face, she departed the building. At 19 Lillian acquired her student pilot license and started working at Darr School as secretary to Herb Glaser and chief pilots Henry Jerger and B.G. DeWitt.

It was not long until a Link instructor married an American cadet. She was quickly fired because policy dictated that no employees were permitted to have any personal contact with the British men. Lillian saw her chance and literally got on her knees, begging Henry Jerger for the job. She became a Link instructor. Her cheerful disposition endeared her to all her students. After the school closed Lillian worked in Kansas City, Missouri and then moved to Dallas, Texas where she worked as a Link instructor for Braniff Airlines. After the war she returned to Ponca City and joined her husband, Harold Taylor, who had just returned from an overseas tour of duty.

In 1965, Lillian, along with her mother and sister, took her first trip to England where she visited Kennilworth. It was because of the combined interest of Lillian and several of the former students, including Bob Phillips, that a major quest began to find as many of the students and staff members of the Darr School as possible. The effort proved to be a monumental task. As instructors and students were found they would fill in lost details about others whom they had known in their classes. Eventually, a comprehensive roster was produced and a series of reunions were held after the No. 6 BFTS Association was formed. Out of the trip to Kenilworth the men of No. 6 BFTS started their association. Their first reunion was held at Runnymede, England in 1976 and Lillian was there. Mrs. Taylor became the major American advocate for the reunion projects and throughout the ensuing years she visited England many times where she helped to

Lillian Bowles poses with one of her Link students on the grounds of the school. *Courtesy Ponca City Library*

Sixty years after the closing of Darr School, Lillian Bowles Taylor was honored with the Pioneer Woman Award for her lifetime work as a goodwill ambassador for No. 6 BFTS. *Courtesy Lillian Bowles Taylor*

renew old friendships between the citizens of Ponca City and the British students.[146] Other reunions followed including some held in England, Canada, and Ponca City. Lillian was honored as Mrs. Oklahoma in 1958 and was presented with the Pioneer Woman of the Year Award in 2003, which was sponsored by the Pioneer Woman Museum in Ponca City. As late as October 2005, Lillian's presence was still noted at the No. 6 BFTS Association reunion held near Oxford, England.[147]

More than 105 flight instructors served on the staff of Darr School of Aeronautics during its time in Ponca City.[148] Of these, four were killed in training accidents with students on board. They were Henry P. Dalrymple, January 27, 1942; Marvin V. Hill, September 30, 1942; J.C. Myers, October 26, 1943; and Alfred August Neumann, January 7, 1944.[149] Henry Jerger was killed June 21, 1943 as he and his mechanic made a maintenance check flight. Jerger urged the mechanic to parachute to safety but he, himself, was not able to escape in time.[150]

Instructors were an eclectic mix of ex-military, crop-sprayers, barnstormers, commercial pilots, and private pilots including Jack Welsch of Woodward, Oklahoma. In 1936, while Welsch was a college student in Alva, Oklahoma, he experienced his first flight. Invited by a friend who was taking flying lessons at the time, Jack took his first lesson. That was all it took for him to decide that flying was what he wanted to do.

He paid for the lessons by working part-time at a drug store. After completing the Civilian Primary Training program, he went to Tulsa, Oklahoma, where he took further lessons in acrobatic flying. Soon after the attack on Pearl Harbor in 1941, Welsch began working at Darr School as a civilian instructor. He taught aerial gunnery, aerobatics, basic flying, instruments, and formation flying. Like most of the other instructors, Jack Welsch had never flown in combat and taught by simulation.[151]

Dr. C.T. Hill, who grew up in Corinth, Mississippi, was no stranger to flying. Roscoe Turner, a famous aerial showman and good friend of Hill's father, partly influenced his desire to fly. While attending the University of Old Miss he visited a fly-in of fighters from the United States Army Air Corp. That fall Hill

The official emblem of the No. 6 British Flying Training School located at Darr School of Aeronautics in Ponca City from 1941 to 1944 was used on letters, Christmas cards, magazines, and photograph albums. The logo still is used today by the association which bears the same name. *Courtesy Paula Denson*

Flying training instructors pose dressed for the frigid temperatures they endured as they taught their students. William B. Kelson is pictured in the center of the trio. Most instructors were not much older than their students. *Courtesy Robert Kelson*

BELOW: Instructors found time for a little horseplay. William Kelson does the honors while another instructor holds down a student who is to be disciplined for "scaring the tar" out of the instructor. *Courtesy Robert Kelson*

ABOVE: Two young training instructors pose in their leather jackets and goggles as they prepare to train a new group of English cadets. *Courtesy Robert Kelson*

began flying and finished his training the next July, at which time he instructed at Bush Field in Georgia. Later, he arrived to continue training pilots in Ponca City from 1941 to 1944.[152]

The youngest of the instructors who came to Darr School of Aeronautics was William B. Kelson, a twenty year old pilot from Scranton, Pennsylvania. Not long after his arrival, he met his future wife, Barbara Ball, who was an employee of the school. Only a year after his arrival in Ponca City, the couple married. Kelson became a career pilot in the United States Air Force after the war. At one period in his career he was a ferry pilot traveling to Alaska in many types of aircraft. He also served in the Korean conflict.[153]

One of the most difficult parts of establishing the new school was the hiring of trained airplane mechanics. Each aircraft used for training would require more than one person to keep it flying perfectly at all times. Mechanics were recruited from the citizens of Ponca City, Miami, and surrounding areas. Anyone who had any experience with automobiles or other engines was offered a job. James Eudell Ledbetter, a mechanic at Darr School, was recruited for a job as he worked for Norris Motors[154] while Jean O'Kelley left a job with Smyer Aircraft Company in order to work at Darr School.[155] Crews worked around the clock in several shifts while maintaining the aircrafts used for flying training. On July 22, 1943, *The Ponca City News* reported the following:

A number of Darr School of Aeronautics pilots, instructors, and mechanics were accepted for the enlisted reserve of the army air forces following physical examinations for induction at the school last week, according to a report from the Tulsa induction station

Barbara Ball and William Kelson were married one year after he came to Ponca City from Scranton, Pennsylvania as an instructor. Kelson met Miss Ball at the school, where she was an employee in the office. *Courtesy Robert Kelson*

James Eudell "Shorty" Ledbetter, a mechanic at Darr School, is pictured on his way to work. Ledbetter was transferred to Miami, Oklahoma in April, 1944 when Darr School was closed for operation. Mr. Ledbetter made his home in Miami for the following eighteen months while he worked at Spartan School of Aeronautics. He remained there until the school was closed in 1945 and returned to Ponca City where he made his permanent home. *Courtesy J.E. Ledbetter*

Dr. C. Thomas Hill, a resident of Ponca City, was a flight instructor at Darr School. After his career as an instructor, Dr. Hill flew as a commercial airline pilot before attending medical school. After his retirement he returned to Ponca City. *Courtesy Paula Denson*

received Thursday by the local draft board.

The induction program was worked out by the war department and the national selective service for all army contract flying schools in order to keep the key personnel of draft age available for the continued operation of the schools.

The men accepted are Harry Wilson Moody, Vernon Earington Roork, George Hayden Whitmore, Clarence T. Hill, Jr.; James Wade Elkins, Isaac Alfred Smith, Joseph Henry Mounger, Jr.; Theodore Milo Quilling, Marion Leroy Backer, and James Hope Hughes.

Carl Dalloe Schuchman, Daniel David Haaz, Frederick Henry Smith, Melvin Frederick Rutz, Robert Maxey Farnam, Ocie Harold Burke, Thom Kittrell McDonald, Farrel Earnest Watson, Robert Lawrence Lucas and Russell Ben Simmons.

Bert Stricklen Henderson, Gerald Wesley Cupp, Wendel Eugene Bowhay, Harley Benjamin Harrolle, John Dee Netherton, William Benjamin Kelson, Charles Edward Johnston, Chester Lesie McCammon, James Wallace Hoyle, and Howard Leroy Haralson.

James D. Golliver, Jack Hubert Evenger, John Wesley Gates, Shelton H. Roy, Robert Francis Adamec, Robert Leon Bass, Vincent Earl Alden, Arthur Elmore Runnels, Jr., James E. Ledbetter, J.D. Black, Elmer Winston Luker, and John C. Aasen.

Kenneth Clapham, Edward C. Goodman, Loyd G. Helm, Ray Eugene Carter, Lewis E. Brown, Clyde William Fellers, John Frank Leonard, Carl Tietz, Jack Wyman Murray, Thomas U. Kelly, Norman E. Hester, Harry C. Conrad, Murray Virgil Mingus, and Earl Crummett.

Joseph Mario Fraccaro, Charles N. Houk, Everett B. Weaver, James Q. Patton, John Cloepfil Merris, Ralph

Donald Schellenberger, Raymond M. Phillips, George Wilmer Welsh, Marion Leroy Stiegelmar, Howard Salyers, and Amret Maurice Taylor.

Arthur John Halbrook, Alfred A. Neumann, Bruno W. Sandberg, Wesley W. Wilda, George Albert Lindberg, James W. Miles, William M. Sturgeon, Euel Irvin Ratliff, Walter Hays Dirham, John F. Kometer, John A. Udove, and Truman F. Chronister, Jr.

Merle Wayne Williams, Leighton D. Smith, Paul Frank Victor, Jr., Raymond H. Carpenter, Joseph F. Frankowitz, Gerhart Max Glaser, Bernard G. DeWitt, Austin Thomas Courson, Francis Wiley Morgan, and Robert Taylor Seamans. [156]

It was quite apparent that there were numerous men who helped the war effort while working in the United States during this time. After the closing of Darr School in 1944, Ledbetter was offered a job as a mechanic in Miami, Oklahoma where he lived with his family for 18 months. He later returned to Ponca City and took up his job as an auto mechanic at Norris Motors where he formerly had been employed. [157] O'Kelley continued as an airplane mechanic, eventually working for American Airlines in Tulsa, Oklahoma. [158]

Elsee Dee Dutton served as a radio dispatcher at Darr School of Aeronautics. *Courtesy Ponca City Library*

A young Ponca City woman, Betty Page, had worked at Mid-American and at the Red Cross rolling bandages before landing a job at Darr School as one of several flight dispatchers. She is pictured next to Doug Londors as he prepares to leave for England. After the war Betty married an American and worked both as a housewife, mother, and in the Conoco Oil Company labs testing aviation gas and toluene. She returned to Ponca City after the death of her husband. Betty continues her friendships with many of the former cadets through frequent telephone calls in which she keeps them up to date about what is happening in Oklahoma. The perky lady keeps a scrapbook of their visits which she shares when former cadets return to Ponca City for Darr School reunions. [159] *Courtesy Betty Page St. Clair*

Mechanics worked day and night to keep the aircraft in top operating condition. Some were recruited from their jobs as automobile mechanics. *Courtesy Ponca City Library*

"Feminine Voices Keep 'Em Flying at Darr School" was the headline in *The Ponca City News* on Wednesday, October 28, 1942 in the following article:

"Women Take Over In Control Tower without Mishaps"

The gentle voices of two women, Miss Elsee Dee Dutton and Mrs. Fredrica Leach, are keeping 'em flying from the control tower at the Darr School of Aeronautics these days.

Since the initial shock when feminine tones broke the stillness of the air like a voice from the heavenly choir, student pilots and instructors are having their ups and downs without mishap under the direction of the new radio operators.

Placing of the young women, under the direction of Lewis Brown, in charge of traffic control, has been an experiment of Henry Jerger, chief flying instructor, as a result of the shortage in male operators. Both now have earned their operators licenses. Miss Dutton formerly attended the College of Emporia in Kansas and has never been up in a plane. Mrs. Leach has taught school three years and has her pilot's license.

The women occupy a little "hot-house" of a glass encased room, the highest point at the school. There is a fine view of the countryside and, incidentally, the flying field.

To the uninitiated, there seems to be a confusion of blue and silver planes taxiing across, nosing off the runway, circling about the field and swooping down on the ground.

In good weather, all practice planes come and go with rapidity, some with instructors and some with only student pilots. On bad days, only advanced ones go up. To reveal their types, if one could, would be telling a military secret. Besides the "gen" (information) which the girls hand out to the pilots about flights, a large flag on the field is changed to indicate types that may fly on any certain day. White does not mean surrender, but what it does mean would be telling, too.

With the number of airplanes coming and going the girls must advise pilots when it is their turn to use the field. Pilots call in occasionally from the towns in Kansas or Oklahoma to which they fly. Weather forecasts are not permitted to be given because they might be intercepted and aid the enemy. However, bad weather may be reported from an individual plane.

The minutes go by, on 24 hour time, rather uneventfully. British accents sometimes cause difficulty but they're nothing compared to the Scots! A true Scottish burr is something to take slowly, they say. Politeness rules always.

Sometimes, as in one case recently, a pilot called in, "Please clear the field. I am making a forced landing. Thank you." Miss Dutton then carried the warning to others through a broadcaster.

Each time a flier wants to start off the field, he calls in his number, the persons in the plane, and destination. He receives permission to "taxi into position and take off when clear." Landing, he is advised "field is clear." On the ground, he moves over to a place near the edge of the runway and receives permission to "switch off engine."

It seems simple, but the operator mustn't become absent minded, as in the case of one pilot about to land. The operator had just told a succession of pilots who had landed to switch off their engines. When a pilot in the air asked for permission to land she said, "Field is clear, switch off engine." However, she corrected her advice before the plane's engine was shut off in mid-air.

The girls say there have been moments when they were serenaded over the wireless, but these have been brief and not encouraged.

Keep 'em flying, girls, you're a better man than I am! --- Abbie Amrine."[160]

Yet another article in *The Ponca City News* on July 4, 1943 stated:

"Women Grease Monkeys Repair Darr Airplanes Third Class for Mechanics is Under Way; Officials say Workers Make Good Showing"

"Hand me a wrench, will you, Jackson? I can't let go of this wing," says one mechanic to another in a Darr School of Aeronautics hangar, so Miss Jackson tucks a stray curl under her bandana and hands over the wreath.

The girls are taking over as grease monkeys at the Darr school, and they "can get just as dirty as the men" according to V.L. (Casey) Jones, school general manager.

Fifteen young women who have completed the 30-day training course are working as mechanics in the 11 p.m. to 7 a.m. shift. The planes are worked over at night so they will be ready to go each morning.

They work under flight chiefs, each of whom heads a crew of eight. A complete check of a plane is divided among the eight according to the sections of the plane, cockpit, tail, wings, landing gear and engine. Each crew member must know how to check whichever part he may be assigned to cover.

Each plane at the field gets a visual inspection after its day's flying and a pre-flight inspection before it is released for flying in the morning. Besides the mechanics who work inside the hangars there are the line crews who work outside, but so far all the girl mechanics are on inside works.

In addition to those checks, a plane has periodic inspections after each 28, 50, and 100 flying hours. In a periodic inspection the parts must be disassembled and checked.

The third class for mechanics started Thursday, and the girls receive $60.00 a month while they are in "school." Their tools are furnished until the second month.

This is a practical school, however. Four hours every afternoon, from the first day on, the "trainees" work in the hangar. Some may be watching one of the men work on an engine that is hanging from a pulley in one corner. He explains to them what he is doing and why.

Others are helping remove a wing that was damaged so another wing can be substituted and the plane sent up again while repairs are underway on the ground.

Hangar practice ends at 6 p.m. and there's an hour for supper. Then class starts and continues for four hours in the classroom at one side of the hangar. The girls have a manual published by the air corps maintenance department, which explains their work. The general public can't examine it, however, because it's a restricted document.

Diagrams of every working part of a plane are hung on the class room walls or may be brought out as they are needed. An engine stands on a table in one corner, ready for the beginners' inspection.

The girls now in class will have their examination at the end of the month, and those who pass will start work July 21 at 11 p.m. The salary jumps to $80.00 a month and they have six days a week instead of five, as in school.

A progress examination will be given them at the end of August and those who are successful will be raised to $100 a month. Another progress exam at the end of October will qualify them for $120 a month, and the final exam at the end of the year will raise them, if they pass to $130.

At the end of a year's service they may take the first class mechanic's exam after which their salary will increase in two steps to $175 a month.

Those in class right now aren't concerned about the money just yet, however. As one put it, "I'm just dying to get my hands on a wrench. That's all I want." [161]

Ponca City was not ready for such a large influx of families moving into the city and had a difficult time providing housing. To help with the shortage, some Ponca City residents offered accommodations in their homes. Barbara Davis, whose father, "Swede" Allard, was one of the cooks at the school, remembers being moved out of her own room so that her parents could provide a room for one of the instructors and his family. She slept on a bed made on the living room divan.[162]

In 1993 Russell Harkness returned to visit Ponca City from his home in Decatur, Alabama. He recalled that he was assigned as chief of supply with Darr School. While he lived in Ponca City, he stayed in the home of Mr. and Mrs. Everett J. Champion on North Fifth Street.

Ponca City's Chamber of Commerce Lends a Hand

From the very beginning, the Ponca City Chamber of Commerce was asked to assist Squadron Leader H.F. Suren in organizing a way in which the students could be issued invitations by local families, all eager to meet the needs of the youth. However, Squadron Leader Suren made quite certain that the public understood that the training schedule for the students was always their first priority. His able secretary was Margaret Gage, a local Ponca City woman. Suren had nothing but praise for her abilities in helping things go smoothly at the school.[163]

Wing Commander Suren was eager for the Ponca City Chamber of Commerce to provide assistance so that personnel from the school could devote their time to their jobs without the added load of handling these schedules. The squadron leader reported that each course would be 28 weeks in duration and that the students would be trained five days a week. If inclement weather caused delays in flights during the week, make-up flights would be assigned on weekends. When the weather did not interfere, the students would be free for social engagements from two o'clock on Saturday until seven o'clock on Sunday evening.[164] Without written permission, the students were never allowed to venture beyond a forty-mile radius.[165]

The interaction between Darr School officials and the members of the Ponca City Chamber of Commerce exhibited a spirit of partnership, even after the first full year of operation.

BRITISH WAR FILM IN LUNCHEON PLANS

Darr acknowledges Honor Shown School by C. of C. in Ponca

A cordial greeting from Harold S. Darr, owner of the Darr School of Aeronautics, to the Chamber of Commerce acknowledged "the honor shown our school by naming November 6 "Darr" Day for the membership meeting of the Chamber of Commerce." Clifford Wetzel, president of the C. of C., reported Thursday.

Darr added that he had been busy in Chicago and Washington on business connected with the school, but that if it were at all possible he would fly to Ponca City in order to attend the Darr day meeting at 1 p.m. Friday at the Jens-Marie Hotel.

If Darr is present, he will give a talk on general subjects connected with the school and its growth here. V.L. (Casey) Jones, administrative director of the school, will be in charge of the program, which also will include talks by Jones, Wing Com. C.A. Ball and others.

The British cadets will furnish entertainment numbers and Jones will show a British War film.

L.R. Northcutt, who was president of the Chamber of Commerce last year when Darr located his school at the Ponca City airport, will preside at the luncheon meeting.

Reservations may be made at the Chamber of Commerce office or by calling 710. The luncheon will be at 1 p.m. Friday.[166]

Soon after their arrival in Ponca City, twelve RAF trainees from Darr School were invited to become tableaux participants at Blaine Park in honor of the Cherokee Strip celebration. A huge Native American ceremonial pow wow was enjoyed by about 8,000 people and the cadets, eyes popping and cameras snapping, captured the moment as five chiefs were introduced and the men were invited to participate in the Ponca Snake Dance. [167]

At another Ponca City ceremony, Squadron Leader H.F. Suren and Flight Lieutenant George Raymond Shea were honored as sub-chiefs of the Kaw Tribe. During the ceremony, in which Chief Little Standing Buffalo and his people were dressed in colorful feathers and beads, the two RAF officers were given rings as a symbol of their membership in the tribe. That same evening, Gus McDonald and Everett Vasser, both Poncas, performed for more than 100 British cadets who witnessed the ceremony and participated in a victory dance. McDonald was once the champion dancer of all North American tribes while Vassar was a talented rope twirler. [168]

The city of Pawhuska invited cadets stationed at both Ponca City and Miami to visit their city during June, 1942. The Pawhuska Chamber of Commerce sponsored the "prayer for victory" dance, which had not been performed since World War I. [169] Because of their study schedule, Darr School students were not able to attend. [170]

Some of the Native Americans and the cadets developed close bonds during their time at Darr School. David Sealey, Course 14, kept a journal recording his

Harold Cogman, pictured at the far left in the middle of men posing for the camera, sits in the grandstands of the Ponca City rodeo grounds as he and his classmates enjoy their first taste of the Wild West. *Courtesy Mrs. Harold Cogman*

Members of the Ponca Indian pow wow pose as a group of Course 9 British cadets capture them in photos. Several cadets were asked to pose with the Indian Princess. *Courtesy John Cal Price*

time in Ponca City. In it he related that at some time while training in Ponca City, he had actually become a blood brother to one of the local Indians and he had the scar to prove it. His wife never learned who David's blood brother was.[171]

It was proposed by the Ponca City Chamber of Commerce that citizens issue an invitation through proper channels and if the students' flight schedules were not behind on Friday afternoon, the committee would issue RSVPs at that time. It is unclear if this proposed handling of social functions ever came to fruition.[172]

As the school in Ponca City began to train the first several courses of students, more barracks and hangars were constructed.[173] By October 14, landscaping around the buildings in Ponca City had been started, with many Chinese elm and American elm shade trees and shrubs planted on the site. There also was a variety of dogwood, Virginia cedars, Japanese quinces, and evergreen bushes.[174]

This is a view of the first hanger constructed at Darr School.
Courtesy Ponca City Library

By April, 1942 a recreation center had been completed and a library, staffed by the Works Progress Administration (WPA), contained more than 350 books donated by the library and by the WPA. Magazines from the dead letter department at the post office were also donated by Mrs. Blanche Lucas, the postmistress.[175] Realizing the need for a place for the students to escape the heat of summer training, a swimming pool was generously built by local oilman and philanthropist, Lew Wentz. A tennis court also was constructed in time for the upcoming summer sports season. Mr. Wentz donated a radio to the recreation room at the school.[176] Many of the young men enjoyed listening to the music of the local radio station, WBBZ.[177]

In 1943, Fredrick B. Bowker recognized that there was a need for public transportation in all areas surrounding Ponca City. The problem was a result of the nation-wide rationing of tires and gasoline. Darr School students and staff were just part of the community which needed this service. Bowker took the initiative and soon he established a Ponca City bus line which operated six days a week. Beginning with a station wagon and three large blue buses, he had several drivers: Fern Freeman, Mr. Charles Shields, Mr. Alvin Grogan and his son, Kenneth, Mr. Floyd Cobb and his wife, Edna, Mark L. Hurd, and Bowker's sons, Robert "Poncho" Bowker and Jack Bowker. The Bowker boys reported that there were times late at night when they picked up groups of cadets who were a little "tipsy" and welcomed a ride back to their billets.[178]

There were three routes: One route covered the area within the city limits including all residential areas, the Nonnamaker addition, Dixie Hill, and "Up town Ponca." Three trips also were made daily to the Ponca Military Academy. The second route made three trips daily to the Ponca City airport and Darr School of Aeronautics and the third route made one trip daily to the Tonkawa Prisoner of War Camp where military prisoners of German and Japanese descent were confined during World War II.[179]

Mr. Bowker maintained this much needed and appreciated bus service from the beginning of 1943 to the end of 1946 when the war was over.

Social Life Was Aided by Local Citizens

By August, 1942 a social center was set up in the home of Mrs. George Brett at 305 South Fifth Street. Mrs. Brett was from one of the founding pioneer families of Ponca City. Her daughter and granddaughters enjoyed entertaining the some of the cadets of Course 17. Equipped with magazines, games, and records, the home provided reading rooms, game rooms, and light refreshments. Wayne Moler, of the city park department, provided 10 park benches for the lawn of the home. Some of the hosts and hostesses who offered assistance during the time the social center was active were Mr. and Mrs. Ed Souligny, Mr. and Mrs. Stuart Clark, Miss Lillian Vance, Miss Mary Dean Vance, Mr. and Mrs. G.P. Broaddus, Mr. and Mrs.

Dave Watson, an unidentified cadet, Floyd Childers, Len Harrison, and Ted Newman pose outside Cuzalina's Drug Store. *Courtesy Ted Newman*

British Students Presence Acknowledged State Wide

The Daily Oklahoman reported that *"---even if the boys aren't at home, they are happy, and the breathlessness with which Ponca City first greeted them has passed. It has become the usual, not the unusual, to hear the drone of many planes in the air day and night, and to hear gray-clad youths, in British accents, call for chocolate malts in the drug stores."*[180] *The newspaper went on to report that there had been no arrests of the gentlemanly youth from England and that although the youth had only rudimentary training in the military at their arrival, they were learning to view their planes as weapons, ready to fight Germans upon their return to England.*[181]

BELOW: A bevy of Oklahoma beauties line up for their English boyfriends. Sixty four years after the school closed, some of the women stay in touch with their friends in England. Pictured left to right are Kathryn Falen, Virginia Sidles, Betty Page, Virginia Drake, Betty Fahs, and Virginia Edwards. *Courtesy John Cal Price*

E.H. Goff, Miss Marilee Barber, Miss Audrey Nelson, Miss Dahlgren Quimby, Miss Joyce Galanter, and Miss Elizabeth Brett. Others who contributed magazines, games, and other equipment were Mrs. H.R. Needham, Mrs. James Hutchins, Mrs. L.D. Edgington, and Miss Genevieve Cummings.[182] Not long afterwards, Miss Ellen Brett had to leave for her work in Chicago and Mrs. Brett announced that there were not enough cadets visiting the center to justify its existence. She believed it was not being used to the fullest because so many invitations to other homes were being offered to the cadets. Thus, the social center was closed after one month.[183]

USO

In Recognition of

MERITORIOUS SERVICE

IN USO WAR WORK

this certificate is presented to

Mrs. W. O. Fuller

CHAIRMAN, USO COUNCIL — PRESIDENT, UNITED SERVICE ORGANIZATIONS, INC.

March 31, 1945

DATE

TOP: The United Service Organization issued this certificate to Mrs. W.O. Fuller for her devoted service to both the men in the United States and British military. *Courtesy Dr. and Mrs. J. E. Fuller*

BOTTOM: Cadet Ted Newman, left, clowns around with his friends, Len Harrison and F.C. Alderman, as they rest near the Civic Center Fountain during one of their furloughs. *Courtesy Ted Newman*

Two Ponca City British cadets enjoy visiting one of the many USO clubs established all over the United States. Many of the men visited them when they made cross-country sight seeing trips during furloughs. *Courtesy Mariann Bryant*

Mr. and Mrs. W.O. Fuller entertained many Darr School students from 1941 to 1944. Mr. Fuller was affectionately known as "Poppy." Pictured are Mr. Fuller, Victor Goddard, and Mrs. Fuller. *Courtesy Dr. and Mrs. J.E. Fuller*

As the need grew locally, a Ponca City chapter of the United Service Organization was founded on October 12, 1943 in order to provide for the support and entertainment of thousands of young men and women in uniform. Chairman of the USO committee was local business man Wilfred B. Johnson, a local banker, along with help from Mrs. Don Gilman and Mrs. Hal P. Leaming, who chaired the junior hostesses.[184] The USO operated in what is now known as the Royalty Building, located on the corner of Grand Avenue and Fourth Street.[185] The young fly-boys spent many hours there writing letters home and visiting with the local girls and boys. Countless hours were spent dancing on Saturday nights and a multitude of other parties also were enjoyed. Matrons played ping pong, cooked for the young men, and acted as seamstresses when the occasion was needed. Lonely young men were assisted by the offer of free long distance telephone calls and telegrams in order that they could visit with their families back in England from time to time. It was at the USO that many were approached by busy homemakers who issued invitations for one or two young men to share a meal or visit church with their families.[186] Mr. and Mrs. W.O. Fuller were one of the couples who took many of the young men home for dinner week after week. Later, they were cited with a letter of appreciation by United Service Organizations, Inc. on April 4, 1945.[187]

More than sixty organizations took part in running the USO during its one and a half years of existence, including groups such as the Twentieth Century Club, Daughters of the American Revolution, veteran's groups, schools, churches, and garden clubs. Louise Fluke, designer of the Oklahoma state flag and an active member of the Ponca City Chapter of DAR, held the title of USO coordinator for DAR during these years.[188] In February, 1942 this chapter reported that 148 British boys, students at the Darr School of Aeronautics, and future pilots in the RAF were entertained in the homes of the members.[189] After the April 1944 closing of Darr School, the USO continued to operate, although its use declined as more and more local servicemen were stationed overseas. In December, 1944 William F. Temple, associate regional executive of the organization, announced that the Ponca City USO would close on April 1, 1945. At that time it was announced that a veteran's center was being planned by a local group to help men as they returned to civilian life.[190]

Wing Commander Charles A. Ball arrived in Ponca City to take command when Suren was assigned to Washington, D.C. in April, 1942.[191] Ball's wife, Irene, and two small daughters, Amanda, 4, and Diane, 9 weeks old, joined him later in the spring as he took his role as head of the staff at Darr School.[192] He had joined the RAF in 1929, serving in the bombing division of the RAF before his assignment to the flight training division. In 1937 he was awarded a Master Flying Instructor Certificate and was an instructor for twelve years. In 1945 he was decorated with the Air Force Cross.[193] His hobbies included swimming, golf, and model yacht building.[194] After the war, Ball, who was an accountant, immigrated to the United States in 1949, settling in Portland, Maine. He died in Cape Elizabeth, Maine on November 11, 1999.[195]

As the United States and Britain hashed out the problem of who should have authority over the men who trained at the schools, there were many mix-ups about who was in charge of what. At the British Flying

Training School in Clewiston, Florida, American Brigadier General J.G. Williams visited the school and performed an inspection of their barracks. He was appalled by conditions which he described as "deplorable." It caused him great concern to see the barracks' rooms looking like "concentrated pig-pens," although he did admit that it was Sunday and the cadets were on their own time. He also was unnerved by the lack of a prompt salute and "drill was not impressive." However, he did admit that the men were on their own time during his inspection. Later, the General finished his report by saying: "...the idea had its good points." History points out that because the training schools were run by the British, the schools' contributions led to a deeper respect for the British manner of training as the Americans joined the scheme. This seemed especially true with regard to the handling of navigational training and to the Brigadier General's lack of attention to the good points of the 'basic training' phase of the flying curriculum." [196] Richard Ravitts, a former USAAF cadet and his wife, Gail, had their own opinion of the general's judgment when they said, "Those Brits didn't understand how hard it is to fly a desk, and they were messy and undisciplined, so how come just a few of them in Spitfires could save their country from a German invasion?" [197] They mentioned that they and their American friends had nothing but respect for their British counterparts.

Cuzalina's Drug Store was one of the most popular hangouts when the British and the Darr School staff came to town. They could always find what they needed on the well-stocked shelves and the soda fountain was always in use. Sergeant Danny Freeman was one of their most regular customers. *Courtesy Ponca City Library*

As soon as Wing Commander Ball arrived in Ponca City, the easy going attitude for rules and military precision which Commander Suren had allowed during his tenure changed drastically. When Richard Ravitts, an American cadet arrived, discipline at the school, much like that observed at Clewiston, Florida, was still very lax. Ravitts reported that it was easy to avoid getting caught for staying out late and that the men began to return to the school whenever they wished, rather than at the hours posted by the school. Dick went on to say:

Then one night disaster struck. The Oklahoma cracker who was the gate guard suddenly had orders to log us in if we were late. "Y'all are late and Wing Commander Ball wants you fellers to rait yer names on this here paper."

Johnny Segelquist and I looked at the list and it was glaringly obvious that the guard could not read, because the names written were Wildcat Grumman, North American Mustang, Lockheed Lightning, P.T. Stearman, A.T. Texan, Martin Marauder, etc. So we added our names, Curtis Warhawk, Bell Airacobra – or something. The next morning, at O dark early, there were 16 of us in front of W.C. Ball's desk and the Wing Commander lowered the boom. He confined us to the base for the whole weekend. [198]

Mrs. Ball, the wing commander's wife, made a hit when she arrived several months after her husband. Mrs. Ball was delighted with the beauty of the city, including the trees and parks. She and the children had traveled to America from West Kirby, England in an American boat carrying 150 women and 72 children. Amanda was very happy to enjoy the plentiful amounts of sweets offered to her upon her arrival. In her short life, she had lived where sugar had been rationed so coming to Oklahoma was a sweet treat to her. However, because of the heat, her blond pigtails were cut off shortly after her arrival. Mrs. Ball was quite impressed with the designer trends in clothing and accessories which were available to her in the local shops. She had missed items such as shoes, pens and pencils, stockings, dresses, and kirby grips, known as bobby pins, due to the rationing in England. [199]

Prior to her arrival in America, the Balls had lived in Hull, England, where during three of those months, the city was hit by a series of bombs. Wing Commander Ball had built a bomb shelter in their garden where the family would sleep at night. Rhubarb was planted on top of it to add camouflage to the shelter. During the day Mrs. Ball tended a mobile canteen, serving bread and soup to persons near her home.[200]

The English Appreciated Their American Friends

In Ponca City, Wing Commander Suren, who was replaced by Wing Commander Ball, always was touched by the kindness and generosity of the Oklahomans and he was constantly amazed when Ponca City families waited at the railroad station to say goodbye to the boys who washed out and were sent back to Canada for training in another branch of the RAF. When he left Ponca City for a new assignment in Washington, D.C., he wrote the people of the city a letter which was published in the local newspaper. It said:

As you know, I have been posted from my work here to Washington. Before leaving, however, I would like to tell you how much I have appreciated the great understanding you and your staff have shown me. In every instance your staff has shown 100 percent cooperation, particularly with reference to the use of items concerning our war activities.

I hardly need to say 'Thank You'-it must be obvious how we all feel about such help.

I would also like, if I may, to express through your newspaper to the people of Ponca City – I am tempted to say, 'My fellow Poncans' – my sincere thanks for all they have done for us. Many remarks have been made to me about the behavior of my boys. No one could possibly fail to try to behave himself when faced with the true friendship and hospitality that all of you have shown to them. I cannot express my appreciation for the way the homes of Ponca City have been opened to us all.

You have made my work here so easy and happy that eight months have gone like a few weeks. My successor is a truly lucky man. That is all, Mr. Editor, and many thanks for this space – may it not be our last collaboration.[201]

Wing Commander Ball was eventually replaced by Alan Rogers.[202] All three commanders worked jointly with the American personnel at No. 6 BFTS during their stay in Ponca City.

The ranking non-commissioned officer in charge of the permanent personnel in Ponca City was stationed in Ponca City for more than two years. Sergeant Danny Freeman, because of his quick wit and friendly manner, was to become a source of daily entertainment for most of the businesses and civilians in Ponca City. Fond of cigarettes, Danny purchased a pack at a local drug store and as he was leaving, the owner called "hurry back." Immediately Freeman returned to the counter, puzzling the owner until he realized that Danny was obeying his command to come back. He told Danny that he had only meant the comment as a good-bye for the Sergeant. Danny replied, "I heard you and I did."[203]

Yanks in the RAF

In Ponca City, United States Army Air Force cadets of 323rd Flying Training detachment entered training with the Britons beginning at Course 5.[204] The USAAF was known as the United States Army Air Corps, or USAAC, prior to the United States' entry into World War II.[205] The American cadets consisted of about twenty percent of each of these courses and were enrolled to satisfy some of the Lend-Lease Act's regulations between the United States and Great Britain. The resulting mix of Americans and British proved to be completely compatible with lasting friendships enduring for many years.

The Americans, almost all of whom had some college credits, were chosen from the Classification Center (SAAC) at San Antonio, Texas, where they were carefully handpicked on the basis of their test scores, time in college, and personalities before being placed in the flying training schools.[206] On the average, they were several years older than their British counterparts and for some reason they experienced a much lower wash-out rate than their British counterparts. However, they did not do as well in ground school studies.[207] Dick Wilma, one of the American cadets sent to Ponca City, pointed out that while he was at the San Antonio Air Cadet Center (SAACC) he, along with 9,700 other young Americans, was interviewed in order to determine whether he was suited to become a navigator, bombardier, or pilot. Since the men were to become part of a good will experimental program, they were chosen because they already had private licenses and were expected to succeed in the program.[208] A friendly

Pictured in full dress uniform are some of the United States Army Air Force cadets stationed at Darr School.
Courtesy Friends of Marland's Grand Home Foundation

rivalry developed between the Americans and their British counterparts as to who were the best pilots. During the time Dick Wilma and his friends were at Ponca City, the seventeen Americans rated from #1 through #17 on flying ability while they rated at the very bottom of the list in ground school studies.[209] The English had them covered on this end of the training. Nonetheless, they were proud to be in the first class to graduate from No. 6 British Flying Training School on May 24, 1943. They were presented with both RAF wings and commissioned as second lieutenants while receiving USAAF wings. That evening, Richard Ravitts, one of the Americans, played piano at the graduation banquet held at the Continental Oil cafeteria.[210]

James Neale Thorne was eager to join the RAF. From an affluent family, he gave up his college studies to learn to fly. He was later killed in the war. He is pictured with his British friend, Albert Sims, who stands just behind him. Years later, Sims, of Enid, enjoyed reading the book documenting the life of the young American flier which was published privately in 1956 by Thorne's family. *Courtesy Albert Sims*

The next day, all of the Americans boarded a vintage train bound for Alpena, Michigan, where they waited for their next assignments. Many were later

Two Stearman PT17 training planes fly over the colorful "patchwork" of Oklahoma fields described by American James Neale Thorne as he wrote to his father in New York. After Pilot Officer Thorne trained in Ponca City, he flew many missions over Germany and was awarded the Distinguished Flying Cross on August 12, 1944. On September 6, 1944, as he flew his 112th mission, Thorne was declared "missing in action" over Arnhem.[219] *Courtesy Albert Sims*

dispatched to the 24th Transport Transitions Training Detachment in Kansas City where they entered Trans World Airline's captain's school. Of the 20 who were sent, thirteen trained in Ponca City, Oklahoma.[211]

One of the Americans who trained in Ponca City and who was later to die in the war was James Neale Thorne. He came from a wealthy New England family. In letters written home to his father, he described Ponca City as a typical small oil city and as a "dull place– two movies, two roadhouses, and prohibition."[212] He further commented that the United States could take notes from the British because of their excellent air corps training program and that as a group the English fellows were "damn fine."[213] Jim was impressed that the Brits went out of their way to be friendly and helpful.[214] While Jim trained in Ponca City, he watched a swimming pool being built at the school[215] and noticed that the local farmers had been able to cut their wheat and bale their hay despite an exceptional amount of rain.[216] Thorne was quite impressed by the colorful patchwork the country provided from his view in the air and he noted that it was quite easy to spot fields suitable for forced landings.[217] Of the complaints which he voiced to his father in his letters, he seemed to be bothered by the Oklahoma heat of 104 degrees and by the fact that Oklahoma was a "dry" state except for beer and bootleggers.[218]

Jimmie Lichlyter, another American cadet who trained in Ponca City on Course 6 but was later eliminated, expressed his reaction to training with the British pilots:

The most memorable day in my life was that one when I entered the gate of No. 6 BFTS. Curious, enthused and excited are mild adjectives compared to the feelings of a stranger moving into strange territory occupied by people with whom he had never been in contact. This feeling was even more enhanced when I was informed of the prohibiting of an American being in RAF uniform. Weeks of dreams of that uniform and the honor of being a recognized member of that organization all Americans believe to be the greatest fighting force in the world seemed to fade.

After being in the camp only a short time and enjoying the comradeship and congeniality of the Englishmen I felt even more ambitious to become one of them and to be recognized as such. Terms and expressions were quite confusing but I should imagine the boys found me a little hard to understand also. — On each passing day I find myself being drawn closer to the friendship and attitude of this organization. I marvel at the pleasantness of its personnel and am proud to be a part of it.

All Americans point with pride and envy to the courage and fortitude of the English, and as a representative of the former it is not only my duty but also my prayer to always be worthy of that commendation. In closing I wish on behalf of this country to say goodbye and sincerest wishes to A and B flights. Best of luck and may we meet again.[220]

Lichlyter, American cadet Bill Barbre, and their wives and friends entertained S.J. "Jerry" Eaton and other British cadets in the Century Room of the Adolphus Hotel in Dallas during one of their holiday breaks.[221]

As the Americans graduated, they were commissioned 2nd Lieutenants in the USAAF and were awarded both American silver wings and the RAF wings.[222] Most went on to assignments in the Air Transport Command as combat fighter pilots or flight instructors and they were considered indispensable to the war effort.[223] Many years after the war, when both schools formed their own associations and began to have reunions, the Americans were right by their sides in all activities.

Most fliers graduated from each course after relentless and exhausting exams but others were eliminated due to bad grades, an inability to control an aircraft, or for a variety of other reasons were "washed out."[224] The standard protocol when this happened was to transfer these men back to Moncton, New Brunswick where they were reassigned for other training, later to be returned to England before assuming other duties. Many became navigators or served in transport positions. Others flew after more training while a few were sent back to their former civil jobs in their homelands.[225]

An aircraft and his pilot met an unhappy end when they crashed during training exercises. Many other cadets who performed less than adequately were "washed out" and returned to Canada without their coveted wings. *Courtesy Ponca City Library*

A North American AT6 Harvard is portrayed crashing between two trees in a Miami's Spartan School of Aeronautics *Open Post Magazine* cartoon. Other cartoons in the magazines displayed student humor regarding flight accidents and training protocol at the school. *Courtesy Ponca City Library*

Twelve years after the death of James Neale Thorne his family restored the Guild Chapel at St. Martin in the Bull Ring, located in Birmingham, England. The church was damaged in a 1941 bombing. *Courtesy Paula Denson*

Cadet Billy Barnes enjoyed visits to Oklahoma City during leave. He kept a list of names and phone numbers of pretty girls whom he met there.
Courtesy Brian Barnes

Ted Newman, standing, and two friends are shown hitchhiking to Tulsa on one of their Saturday leaves. The men never had any trouble thumbing a ride if they were dressed in uniform.
Courtesy Ted Newman

Some cadets were reassigned (washed out) after pulling unauthorized flying stunts. Squadron Leader G.S. Puddy, Darr Course 13, related his "wash out" experience:

I commenced flying training at Ponca City on the 31st May, 1943 with a Mr. Mearl Williams as my instructor. Unfortunately my stay was to be but a very short one. On the 24th July 1943 I was sadly led astray by 3 wild USAF cadets and one crazy RAF cadet, and we tried a re-run of a 1914/18 dog fight, starting at some 5000 feet in our little PT17's and ending up at (I shudder to think) over a little town called Kaw City. We were, of course, caught – a dual aircraft in the area spotted the shambles taking place over Kaw City, joined in and took our numbers. On landing we were subjected to what can only be described as 'and verily the sky did fall on them from a very great height'. The other RAF cadet, a Flight Sergeant W. M'Cabe (WOP/AG) DFM and I were rushed with indecent haste back up to Canada under Open Arrest, whilst our three USAF friends were quickly spirited away to a home for bad boys and wayward sons.

It took some two months in Canada before it was all sorted out. M'Cabe, being a 'regular', was worldly wise and when asked by 6 BFTS whether we wished the evidence against us to be recorded on oath replied that it was unnecessary – made no comment when all the written evidence referred to our crime being "rat-chasing" as opposed to "dog-fighting" – the equivalent RAF expression; "rat-chasing" thus being an unknown pastime in the RAF, and certainly not one to be undertaken by cadets hoping to become "Officers and Gentlemen." Consequently, the charges when presented to the RAF Court in Canada just did not stand up – although we were quite obviously as guilty as Hades. I also think Mac's having completed two operational tours in Bomber Command as a WOP/AG and being awarded the Distinguished Flying Medal had a not inconsiderable bearing on the Court's approach to both us and the trial. Incidentally, as a WOP/AG Mac' played a major part in the wartime film "One of our Aircraft is Missing." [226]

Joe Moyle, Course 16 at Darr School, reminisced about one of his temptations while training in Ponca City. Joe and another buddy decided they were going to fly under the Old Arkansas River Bridge but the closer they got, the lower the bridge appeared and they changed their minds before they tried the foolish stunt.[227] If they had not been killed or injured, they were at least certain that they would have been eliminated post haste. It was this type of crazy stunts which did send several young men back to Moncton without their wings.

Weekend furlough was one of the high lights of the cadet's stay in Ponca City and in Miami. They journeyed to the bright lights of the big cities as fast as they could hit the road, hitchhiking most of the time. Billy Barnes, Course 2 at Ponca City, kept a small date card among his personal mementos gathered during his time in Oklahoma. The card, an advertisement for Kelley's Pine Inn located 9 miles east of the Oklahoma State Capitol on Northeast 23rd Street, Highway

62/270 in Oklahoma City, contained names and phone numbers organized under headings of "Blondes" and "Brunettes." "We're JUST Catty Enough To Keep the Rats Away" was the Pine Inn's motto. The card also advertised excellent food and a perfect floor for dancing.[228] This popular meeting spot was located very near the new Tinker Air Force Base and it seems likely that some of the cadets, including Billy Barnes, ventured there on leave to see the huge bombers based there.

Jack Boone and Peter Dowthwaite, Course 17, made friends with Mr. Jack Clarke when he picked them up on North Union Street as they hitchhiked to town. Mr. Clarke's father had come from England and so he felt a connection to the friendly young men. They were invited to Sunday dinner almost every week and it was soon evident that they loved walking after their meals. Mr. Clarke would join them as they took a jaunt from Comanche Street, near the Ponca City Power Plant, proceeding east on Highland Avenue and continuing along Lake Road. After walking about three miles east they would turn north and walk uphill to Prospect Avenue before turning west as they made the final long walk back to Comanche Street. All in all, their brisk walks covered a loop of six to eight miles and were not for the faint of heart.[229]

At the end of the students' training, ground school examinations were given.[230] The men were also required to conduct long distance cross-country flights with two men per plane prior to graduation. They would alternate their flying with one acting as pilot while the other was the navigator. Some teams flew all the way to El Paso, Texas, and back.[231] Typically, cross-country flights to destinations from Ponca City would be Wellington, Kansas, El Dorado, Kansas, Winfield, Kansas and Stillwater, Oklahoma.[232]

Cadets Venture Far and Wide

After Primary training and sometimes following graduation examinations, the men were granted seven to ten days of leave.[233] Cadets from both Ponca City and Miami took advantage of the time by traveling to far away places such as Chicago, New Orleans, Fort Worth, Denver, and the Grand Canyon. One or two were said to have made it as far as Hollywood. Some of the cadets used their servicemen passes for discounted tickets on railroads and buses while others hitch-hiked their way across country.[234] The Ponca City Chamber of Commerce also assisted when they put out a request in the newspaper asking for persons willing to offer transportation for leaves to please call their office.[235]

Bert Sims' leave took him to Colorado where he enjoyed the panoramic view offered from Lookout Mountain. *Courtesy Albert Sims*

Mrs. Roosevelt had a special friend whom she had asked to be an unofficial hostess in both Ponca City and Miami. She was a Kansas City, Missouri woman by the name of Lucille McPherson Jenkins. The wife of Paul W. Jenkins, of the Jenkins Music Company, was affectionately known as "Mama Jenks" by the young men who visited her home during some of their free time away from training. Mrs. Jenkins and her family lived in a very affluent home at 1023 West 55th Street in Kansas City. Cadet Peter W. Smith and a host of other cadets from Ponca City and Miami sometimes visited the home during their leave.[236] On one of her visits to Ponca City, "Mama Jenks" and her daughter, Lucille, together with Miss Elizabeth Ess, gave a party for the cadets at the Jens-Marie Hotel.[237]

British visitors to Mama Jenks' home signed their names, service number, and the date in a brown leather guest book, later owned by Alta Dixon, one of the Ponca City hostesses. Many offered expressions of thanks as they signed their page. On December 6, 1943, Sgt. Robert G. Woodwards of Course 14 wrote:

This house and those in it holds a section of my heart for ever. There is no place like it – there are no people like you. You are tops and I will never forget you.[238]

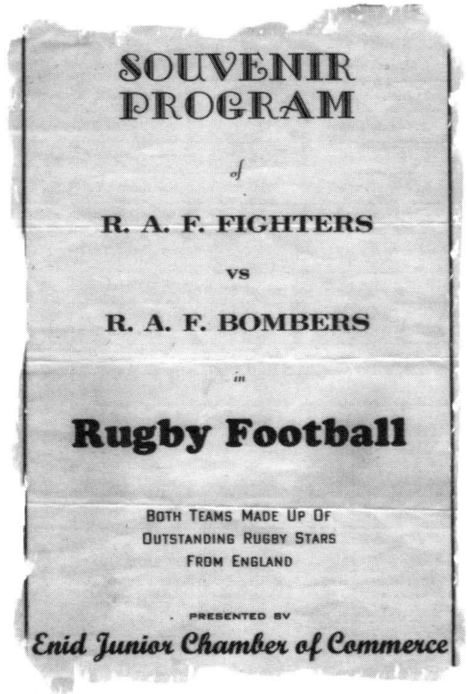
SOUVENIR PROGRAM
of
R. A. F. FIGHTERS
vs
R. A. F. BOMBERS
in
Rugby Football
Both Teams Made Up Of
Outstanding Rugby Stars
From England
Presented by
Enid Junior Chamber of Commerce

Soccer and rugby were popular sports brought to the United States by the British cadets. They routinely held exhibition games in Ponca City, Miami, and many other towns in Kansas and Oklahoma.
Courtesy Ponca City Library

Dennis Stone, left, and John Cal Price as they arrived in Colorado Springs during leave. The two men enjoyed riding horses during their stay in the area. Later, both men wrote letters to Mrs. Verba Hocker, one of their Ponca City hostesses, after their return to England.
Courtesy John Cal Price

Calisthenics was an important part of every cadets training. The future pilots needed to be in excellent physical condition as they headed back to fight the Nazi regime. *Courtesy Ponca City Library*

Another guest, Sgt. Pilot G.T. Gough, echoed his sentiments by writing:

There is no greater compliment that I could possibly pay than by saying that when I am with you in your home I am reminded of my own home.

In October 2005, Bill Neilson, a Ponca City cadet, recalled that some sixty odd years earlier, he and his friends arrived at the USO in Kansas City for a holiday and they were immediately told that Mama Jenks would be along in just a little while. Sure enough, she took the men to her home where they were treated like royalty.[239]

Let's Play Ball

Sports, such as boxing, soccer, rugby, and cricket played a very important role in the physical training and endurance of the cadets. It was something they loved to do. Rugby, a game "involving most of the thrills of American football, some of the finesses of hockey and a little of the technique of basketball" was played with fifteen players on a side. Two forty-minute halves with no rest periods were played. The contenders on both sides wore no padding or guards and wore shoes similar to those used by American teams. Substitutions were not allowed after the game began.[240]

Several rugby games were exhibited with Ponca City teams playing at Taft Stadium in Oklahoma City. Proceeds from the matches, paring the English and Welsh against the Scotch and Irish in friendly rivalry, were placed in the RAF benevolent fund. Soccer teams require 11 players on the field per team while the rougher rugby used 15. General admission was 50 cents while reserved seats were one dollar.[241] An anonymous donor paid $5,000 for the ball which was autographed by all 30 players. This money was earmarked to buy War Bonds.[242]

One team of cadets from No. 6 BFTS traveled to a scratch rugby game in Miami, Oklahoma. They lost the rugger match 41-0. Traveling on a school bus designed

Lucille McPherson Jenkins was well known by the cadets in Ponca City and in Miami, Oklahoma as "Mama Jenks." The engaging hostess and her family often entertained cadets in her large home in Kansas City. A friend of Eleanor Roosevelt, she visited both of the Oklahoma towns while the young cadets trained at the schools. Mama Jenks visited the home of Alta Dixon in Ponca City. Both women and their families became close friends with many of the British cadets. *Courtesy John Cal Price*

to carry school children over short distances, the 320-mile round trip was long and weary and the dirt roads were in great need of grading.[243]

The Ponca City Board of Education, interested in the welfare of their British guests, provided the use of Blaine Park for a football game by Darr School students after deciding in January, 1942 that the date of the game must precede the plowing of the field that spring.[244]

Vince Hatchard, a twenty-year-old cadet from New Zealand, was the welterweight champion of New Zealand in 1938 and while he served in Ponca City, he participated in a fight at a boxing tournament for navy relief held in Blackwell.[245] He previously had defeated the British army welterweight for that title in 1941.[246] Hatchard was later killed in action.

David Riekie, Course 11, trained in Ponca City from November 13, 1942 through May 25, 1943. Riekie related that players from No. 6 BFTS put on an exhibition rugby match for the people in Anthony, Kansas, a town located about 70 miles from Ponca City. Accustomed to softer playing fields than those found in the Oklahoma and Kansas grounds, the football field was watered down before the big game to accommodate the players. Attired in attractive short-skirted costumes and white leather marching boots, the Anthony Pirettes,

In July 1942 Cyril Baker and Bill Phillips enjoyed a relaxing swim at Wentz Pool. *Courtesy John Cal Price*

The all-girl 1942 Anthony, Kansas Drum and Bugle Corp performed at half time during an exhibition soccer match presented by some members of No. 6 BFTS as they traveled to the small Kansas town. *Courtesy Anthony, Kansas Chamber of Commerce*

BELOW: Soccer exhibitions were popular while Darr School was training British students during World War II. Local organizations were happy to sponsor the events. *Courtesy Eddie Pugh*

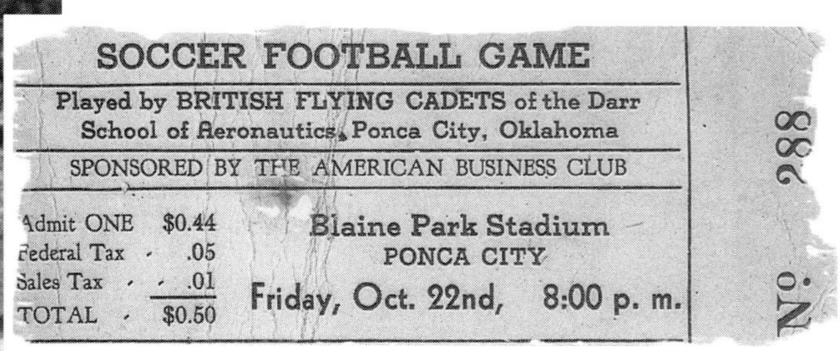

SOCCER FOOTBALL GAME

Played by BRITISH FLYING CADETS of the Darr School of Aeronautics, Ponca City, Oklahoma

SPONSORED BY THE AMERICAN BUSINESS CLUB

Admit ONE	$0.44
Federal Tax	.05
Sales Tax	.01
TOTAL	$0.50

Blaine Park Stadium
PONCA CITY
Friday, Oct. 22nd, 8:00 p. m.

No. 288

LEFT: Swimming and picnicking at Wentz pool were always popular activities as the air cadets enjoyed hot Oklahoma summer days. *Courtesy Sarah Barrington Westerheide*

members of the Anthony, Kansas High School Drum and Bugle Corps of 1942, Riekie remembers they were less than happy as they marched off the muddy field following their half-time performance.[247, 248]

His course was not the only one to visit Anthony, Kansas. In his diary and log book, John C. Price, Ponca City Course 17, wrote that on January 31, 1944 in his Stearman PT17 No. 220, he made a cross-country flight to Anthony, then down to Nash, Oklahoma, and back to Ponca City. The flight took 1 hour 47 minutes and was intended to be flown at 1,500 feet. But, as it was a clear and very cold day, John climbed his altitude to 3,500 feet into a warmer band of air. Just before reaching Nash Price noted that he passed an area marked on the map as a dry lake, but, due to recent rains, was partly filled with water.[249] What John Price observed was the Great Salt Plains Lake.

Let the Good Times Roll

Before all the courses were passed and all the flying hours were fulfilled, the cadets also managed to enjoy their surroundings. During their stay in Ponca City, there was never a dull moment for the young cadets. Everything was new and exciting and they were enchanted with their surroundings. Fond memories brought back old times for Victor Howe of Course 3 when he remembered the new paint on the "huts" in which the young men lived and the very country town atmosphere of Ponca City, where he could cross the road quite casually without bothering too much with traffic. Victor enjoyed riding horses and made note that he and all his friends bought Texas belts of which they were very proud. A silver dollar was kept over the years as another special memento.[250]

On most Wednesday afternoons the handsome young men, sharply dressed in their uniforms, were allowed off campus and "high-tailed" it straight to town. The local ladies, mostly of high school age, were enchanted with the idea of developing friendships with the cadets–after all, they were quite handsome young men in their uniforms and their accents and manners were charming. Moreover, many young American men were joining the military forces and would soon be in scarce supply. Meeting their dates in front of the Civic Center fountain on Grand Avenue became a popular pastime for both the young women and men. Across the street, the men also enjoyed the books and art offered by the Ponca City Library.

Hugh Collins, of Ponca City Course 1, was very interested in meeting the local young ladies and was quite surprised to see girls of 15 or 16 driving "Chevs" down Grand Avenue. Hugh and his buddies happily exchanged photographs with them. After his return to England, he flew Hurricanes and later was assigned to Typhoons with rockets. He was shot down on a low level attack of a power station in France and found himself a prisoner of war. Eventually, he survived the "Great Escape" where 50 out of 87 who escaped the tunnel were shot.[251]

While Bill Barnes trained with Course 2 in Ponca City during 1941, he was befriended by the D.L. Murray family. The Murrays had two boys and two girls, including an eight year old daughter named Karen. After Bill's return to England he continued to

David Riekie, a member of Course 11 at No. 6 BFTS, spent happy hours visiting in the home of Mr. and Mrs. W.O. Fuller *Courtesy Dr. and Mrs. J.E. Fuller*

BELOW: Two happy cadets as they visit with friends in Ponca City. *Courtesy Dr. and Mrs. J.E. Fuller*

RIGHT: The Civic Center fountain, built as a memorial to veterans of World War I by the Ponca City Chapter of the Daughters of the American Revolution, and lit by colored lights each evening, was a favorite spot used by many of the young cadets as a meeting place. David Watson poses for his friend, Ted Newman. *Courtesy Ted Newman*

write to the family and this friendship lasted for many years. On April 20, 1943 Mr. Murray took time to write to Bill from his home at 720 Marland Drive in Ponca City: [252]

BELOW: Pictured are Pete Chattin, left, and Ray Thorogood. Pete was killed-in-action in September, 1944 and he is buried near Tangres, Belgium. Ray died in an accident in England on July 19, 1943.

Courtesy Dr. and Mrs. J. E. Fuller

Dear Bill.

We received your letter of March 24, 1943 this morning. I guess that Jerry has been playing hell with our mail for I have written you three letters since Xmas. The last letter I received from you was posted just after Xmas-I believe on December 28, 1942 and I received it on February 26, 1943 and answered it at once.

The noon radio news is just telling of large RAF attacks on Germany and Italy. From the number of attacks they have carried out on Germany I would think that there would not be anything left to bomb but after all, I guess that Germany is a large country and has lots to bomb. One thing we can say is that at the present time the Germans in North Africa are catching "Hell" coming and going. I hope that by the time you get this letter that the Jerries in N.A. will all be drowned in the Mediterranean and the combined army making their way "thru" Italy.

I told you in my last letter that "Ned" Crowley and Jack Sutherland were united and attached to a squadron somewhere in England. Haven't heard from Bert since he left Canada but still hope to someday, for old Bert was a pretty good boy and we all liked him very much.

I was just up town talking to Dad. He is getting along just fine. Mother is in Texas at the present time since her mother died last week and she went down. She will be home again this weekend. Dad sent her a telegram this a.m. telling her that he would meet her in Fort Worth Friday at noon.

We have all made it "thru" the winter OK with no great amount of illness. Jack at the present has "strep" throat and is just about over it. He has only been sick a couple of days and will be ready to go to school again in a day or two. Bill and Janice are both well and getting fat as little pigs. Karen is still Daddies girl and every now and then asks when the British boys are coming back - says she wants to see Bill, Ned, Jack and Bert.

You say you don't see how we get 15 miles to a gallon on our large cars over here. I will tell you that most of our 6 and 8 cylinder cars get close to 15 miles to a gallon of gasoline - it all depends on how good you treat your car and what kind of a shape it is in. We can only drive 35 miles an hour and that helps a lot to increase the mileage per gallon. You take our large trucks- they have cut down their speed and the load they carry in order to conserve their tyres. It is not a shortage of gasoline that we have but a shortage of rubber to replace the worn out tyres so they just ration gasoline in order to save the rubber on tyres. It is quite a joke, this gasoline ration, here in the Middle West. We don't see why we have to do without anything here in the Middle West of the USA, just because the "rich" on the East Coast can't get theirs due to shipping facility. All we have to do is go to the ration board and tell them we want so much gas and it is ours - to hell with this 4 gallons a week when you can get more just for the asking.

Thanks for the English coins. I will enclose one of our new steel pennies. When you were here our penny or 1 cent coin was made of copper but now the new ones are made of steel. I will send an old one and a new one.

Well Bill, will close -
Your friend DL

This is a typical scene looking east along Grand Avenue in Ponca City during World War II. *Courtesy Ponca City Library*

She was so impressed with the pretty name of the little Oklahoma girl that Bill's aunt named her own daughter Karen. Bill Barnes mentioned in one of his letters to his sweetheart, Eileen Hiley, that the country was lovely and the people helpful.[253] Bill also brought home a letter from the Ponca City Lion's Club which was illustrated with a picture of the famous Pioneer Woman statue. It read:

Greetings: We have cherished your short stay with us so much that we desire to continue this precious fellowship by wishing you great joy for Christmas and the happy experience of peace in the New Year.

A favorite Ponca City hangout was T.J. Cuzalina's Drug Store where hungry young men enjoyed wonderful ice cream concoctions loaded with lots of whipped cream, nuts, and syrup. Vincent H. Twomey, of Course 1, maintained contact with T.J. for some time following the war, never forgetting his kindness. Mr. Cuzalina, who was later credited with inventing the phrase, "I Like Ike," [254] presented Vincent with a silver ring engraved with the words: "Forget me knot. T.J." Twomey flew Spitfires upon his return to England and "ended up as an unwilling guest of Adolph Hitler for the last eleven months of the war- in short, a P.O.W." [255]

In those days alcohol was taboo. However, the men occasionally visited the local honky-tonks to enjoy a beer or two and always seemed to find a way to obtain liquor when they wanted it. On one particular evening, Twomey and another cadet met a fellow in T.J. Cuzalina's Drug Store who later took them horse back riding. Three horses were hired from Mr. Wetump. "Tumps," as he was affectionately called, had a riding establishment which many of the men enjoyed during their Ponca City stay. Twomey and his friend rode to a place called the "Rendezvous" – a honky-tonk. Although Vincent and his friend drank very little, their host did. On the return ride home they rode on either side of him in order to keep him from falling off his horse.[256]

Cadets A. Snell, T. Grynkiewicz, D. Thomas and N. Maclennan visited Wetump's Stables every two weeks, immediately after pay day. Because they had no experience riding horses, some of them were prone to use cushions after the ride. On one particular ride, one of the cadets 'ground looped' man's best friend. He held too much 'back pressure on the stick at take-off'. His horse 'stalled' and 'not having enough altitude to recover' he made a very un-dignified forced landing in a ditch, with his noble steed as company. Undaunted by this incident, he again mounted and surprised them all by staying on for the rest of the outing.[257] As each course arrived, the demand for Mr. Wetump's horses increased so much that "Tumps" had to acquire more steeds.[258]

The British lads took a liking to *Lil Abner*, *Henry*, and *Donald Duck*. Although it was difficult to find much the boys did not like in America, they were not fond of *Dick Tracy* and other gangster comics and thought that American tea was much too weak. Cornbread seemed strange to their palate and they did not much care for the taste of American beer.[260]

Precious souvenirs found their way to England including chocolates, cameras, post cards, and cowboy boots. Many cadets also bought quantities of silk stockings for their moms, sisters, and girlfriends back home.[261]

The English loved American automobiles and everyone roared when the British chaps told their American friends that the Americans drove on the wrong side of the road. Erma Burleson Carmack, a young nurse who was in training at St. Joseph Hospital, laughingly recalled her observations of some of the young pilots of the RAF, most of whom had never driven a car until reaching Oklahoma. They were teaching a local blind man, Henry Barrick, how to drive an automobile. While Henry steered, the boys would

Cuzalina proudly supported the British contingency of young fliers throughout the war and was always one of the merchants who took out a monthly advertisement in the *Tails Up Magazine*. This cartoon is symbolic of the popularity of the soda fountain featured inside the store.[259] *Courtesy Ponca City Library*

Cuzalina's Drug Store was probably the favorite gathering place for most of the No. 6 BFTS cadets. It was located on the northeast corner of Second Street and Grand Avenue in downtown Ponca City. Although the building remains, it is unoccupied. *Courtesy Ponca City Library*

shout directions as to which way to turn or when to brake.[262]

Visiting the cinema on Saturday afternoons was another popular activity enjoyed by the cadets in Ponca City and Miami. The boys in Ponca City had three movie theaters to choose from – the Roxie, the Murray, and the Poncan Theatre. In Ponca City, a late night show was offered at midnight every Saturday night.[263] One of the training pilots in Ponca City Course 5 knew almost first-hand about the mechanics of movie making. He was J. D. Rathbone, brother of Cecil Rathbone.[264]

Dancing was popular and it was never difficult for the good natured, handsome young men to find dance partners. Some tried square dancing but most enjoyed the more traditional steps of the fox trot or the quick step, whereas they observed that Americans simply enjoyed the informal shuffling and swaying to the music. The English boys soon discovered that things were less formal in America and that women thought nothing of wearing slacks to a dance.[265]

Course One cadets hosted their first public social event at a dinner and dance in January, 1942 at the Continental Associates Cafeteria. The event marked their graduation. Yellow and bronze chrysanthemums flanked the dinner tables where 108 guests enjoyed their meal. A dance was held after the dinner with approximately 300 guests. The Range Riders of Oklahoma A&M College in Stillwater provided the music.[266]

The Club Lido, owned by Jack Ernest,[267] did a booming business with the cadets, who were often found there dancing.[268] The club was open every night except Monday and featured an orchestra on Saturday night from 9:00 p.m. to 2:00 a.m. The sum of fifty-six cents got the boys in the door.[269]

Some organized dances, often held at the Club Lido, were sponsored by the student pilots with proceeds going to the benevolent fund for the school. Funds were used to obtain sports equipment and other necessities for the barracks. On one occasion the music was provided by the Southwest Serenaders, directed by Charlie Groat.[270]

Darr cadets and those from Spartan School of Aeronautics in Miami were often asked to special parties in their honor in both Oklahoma City and Tulsa. These were usually hosted by some of the wealthy oil families in those cities. David Pearce, Course 5 at Ponca City, was lucky enough to be included in one of these parties. Bussed to the city, the men were greeted by some of Oklahoma's oil barons who took them by pairs to their homes for the weekend. There they lived in the "lap of luxury." He noted that at the evening dance, the woman were dressed and coiffed beautifully. The bourbon flowed and the men drank their share, although not accustomed to drinking in the quantities which were offered up that weekend. Ultimately, David's roommate paid the price and suffered a tremendous hangover.[271]

RIGHT: Student nurses from Ponca City Hospital held a benefit dance to aid their nursing association in 1943. The dance was held at the Jens-Marie Hotel Grill. *Courtesy Paula Denson*

Many dances took place in Ponca City and in Miami during the time the English were in training. Some of the graduation parties featured a meal with a dance following. *Courtesy Mariann Bryant*

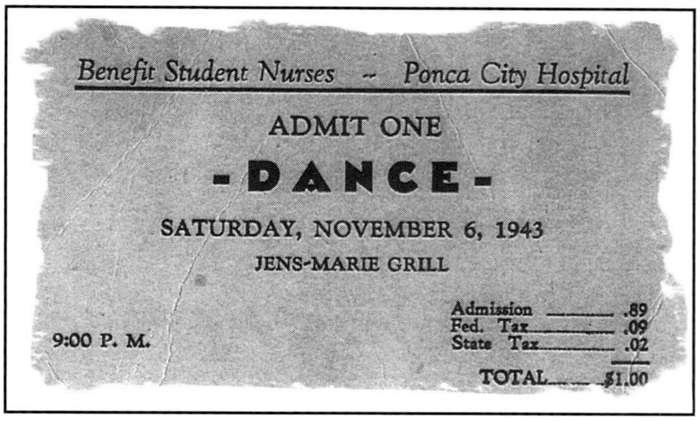

Benefit Student Nurses -- Ponca City Hospital

ADMIT ONE

-DANCE-

SATURDAY, NOVEMBER 6, 1943

JENS-MARIE GRILL

9:00 P. M.

Admission	.89
Fed. Tax	.09
State Tax	.02
TOTAL	$1.00

Jack Ernest, owner of the Club Lido, often placed an advertisement in the *Tails Up Magazine*, published monthly by various courses during their stay in Ponca City. *Courtesy Ponca City Library*

The Club Lido was a popular night spot in Ponca City, no matter which evening of the week it was open. *Courtesy Bret Carter*

The Daily Oklahoman reported the party:

Benefit Dance Will Be Given To Honor Cadets
Bundles for America, Inc. Will Sponsor Affair Saturday Night

British Royal Air Force cadets from the Darr School of Aeronautics at Ponca City will be entertained at a benefit dance from 9 p.m. until midnight Saturday at the Oklahoma City Golf and Country Club.

The dance will be given by Bundles for America, Inc. and will be open to the public. All proceeds will be used to purchase comfort kits for the men in the armed forces of America.

The Ponca City cadets will be guests overnight in the homes of Oklahoma Citians and will also be guests at dinner parties Saturday evening and at breakfasts Sunday morning.

Plans for the affair were made Friday morning at a meeting held at the home of the president of the organization, Mrs. Hugh M. Johnson, 420 Northwest Fourteenth Street.

Members were appointed to serve on the reception, transportation, housing, decorations, music, refreshment and ticket committees.

By Friday afternoon, calls had been received from the following persons wishing to entertain cadets as guest in their homes:

Mrs. E. Guy Owens, Mrs. Frank P. Johnson, Mrs. Harrison M. Smith, Mrs. John J. Harden, Mrs. Hugh M. Johnson, Mrs. Joseph F. Rumsey, Mrs. B. Franklin Gill, Mrs. Stephen Chandler, Mrs. Jerry B. Sass, Mrs. Will S. Bulkley.

Mrs. John W. Teter, Mrs. Robert M. Rainey, Mrs. Edward H. Miller, Mrs. B.V. Whitfield, Mrs. Robert A. Hefner, Mrs. J.W. Upsher, Mrs. Jules Bloch, Mrs. George D. Key, Mrs. Leslie M. Westfall, Mrs. F. Holly Prime, Mrs. Robert U. Patterson, Mrs. Barton Law, Mrs. Lewis J. Moorman, Mrs. W. J. Wallace, Mrs. Frank A. Chilson, Mrs. V.E. McInnis, Mrs. Walter A. Lybrand, Mrs. G.A. Nichols and Mrs. Philip Pierce.

And by Friday afternoon, the following hostesses had asked to entertain at small dinners or breakfasts:

Miss Glennes Teter and Miss Dorothea Rainey, Mrs. Frank P. Johnson and Mrs. Smith, Mrs. Hugh M. Johnson and Mrs. Harden, Mrs. Gill and Mrs. Chandler, Mrs. Whitfield and Mrs. Louise Lane, Mrs. Rumsey and Mrs. Richard D. Rumsey, Mrs. Law, Mrs. Bulkley, Mrs. Sass, Mrs. Miller, Mrs. Owens and Mrs. Charles Anderson.

The dance will take place the same night as the Men's Dinner Club's annual "Ladies' Night" banquet. The two events will not interfere as members of the club and their guests plan to attend the dance after hearing Vincent Sheean, guest speaker, give his after dinner lecture on "The War in the Far East." [272]

Local Activities Keep the Cadets Busy

The British chaps presented several talent shows at the Ponca City auditorium. On July 3, 1943, No. 6 BFTS presented a show titled "Grand Vaudeville." Six gremlins wearing padded flying suits struggled to open the curtain and bounced around getting the stage in order. For the admission of fifty cents,[273] citizens were serenaded with songs accompanied by two American cadets who played the guitar and clarinet.[274] One of the cadets, a professional musician, had his own orchestra in the east before joining the service. American cadet, Vaughn "Monty" Montgomery, belted out many tunes with the voice that people listening to radio station WBBZ had become used to hearing every Sunday.[275, 276]

Stan Ward, as the "Belle of Ponca," also mesmerized the audience as he appeared dressed in women's clothing and a blond wig. When the blonde curls got caught in Michael Cox's coat buttons, they brought down the house with their efforts to become untangled. [277]

Other vaudeville programs were offered by various courses who attended the school. Bryan Bruce-Gardner, Course 15, participated in a charity show held in the high school auditorium while he was at Darr School. In a 1984 letter he commented that he and another cadet, David Weston, had performed an act in which the first verse went:

Roosevelt is boss in the USA but only at times when Eleanor is away. Good old Eleanor, cads, good old Eleanor.

Because of the political views in Oklahoma at the time, the men got much applause. Each verse ended with:

But what of the crimes of (CO) Rogers and Phillips and Lupton and Grime. (RAF officers stationed at Darr School)

Years later, Bruce-Gardner also could recite the Lucky Strike radio commercial played on radio station WBBZ.[278]

The Reverend Gordon V. Smith, acting as chaplain for Darr School, regularly saw many of the English cadets attend the Episcopal Church in Ponca City. In England, about three-fourths of the men were members of the Church of England. Reverend Smith's church took an active role in offering the proper type of recreation and entertainment for the chaps during their stay in Ponca City, as did those of many other churches in Ponca City and Miami.[279]

The Poncan Theatre, located in the 100 block of East Grand, was a popular attraction for British cadets as they watched Hollywood in action. *Courtesy Bret Carter*

Young trainees at both schools were eager to give back to their new communities. V.L. "Casey" Jones, administrative director of Darr School, was in charge of a program presented to the Ponca City Chamber of Commerce in which some British cadets furnished entertainment numbers. After a cordial greeting from Harold S. Darr, who thanked the chamber of commerce for setting aside November 6 as "Darr Day" at their membership meeting, the cadets entertained with songs and Jones showed a British war film.[280]

One hundred members and guests of the Ponca City Chapter of the Daughters of the American Revolution were entertained in the library clubroom at one of their meetings by Arthur Williams, singer, and Joe Zanzottera, pianist–both English cadets. They presented several numbers including *This is Worth Fighting For, George Washington Didn't Tell a Lie, Always You, Danny Boy, Last Time I Saw Paris*, and *Trees*. Refreshments were served at the end of the evening with red zinnias of many shades centering the refreshment table.[281] Zanzottera was later killed in action.

Lewis Ashman, of Course 17 in Ponca City, related his Christmas experience in Ponca City:

Home is at the heart of Christmas, so it may seem strange that the two most vivid recollections of Christmas past that I recall are those spent far away from home.

As a young airman in my early 20s, I found myself stationed at Ponca City, Oklahoma. Invitations flooded in to the Station Adjutant for British lads to stay in American homes over the Christmas period of 1943. Not one British lad stayed on camp and I was sped 40 miles east to Pawhuska, an Osage Indian Tribe settlement.

Neon lights of welcome flashed everywhere as we entered the Episcopalian Church for the midnight candlelit service. The familiar carols brought a warmth and glow of home, heightened at one point when a chorister's surplice caught the flame of a candle, which fortunately swift attention extinguished.

Our British accents were a source of great fascination and amusement to the girls who accompanied us throughout the wonderful round of hospitality and entertainment lavished on us by our American hosts.

What a contrast was the following Christmas of 1944, when having gained my navigator's wings in Canada, we headed for home aboard the "Louis Pasteur" en route from Halifax, Nova Scotia, to Liverpool, across the dark and wintry Atlantic.

Slung amid the hammocks beneath decks, it was the pilots and navigators who received orders to be on watch, 2 hours on, 4 hours off – along side the guns. So that it was on Christmas night that 1805790 Sgt. Ashman found himself on duty beneath the stars – never to be forgotten.

How could I let such a night go by without caroling to the heavens? I somehow discovered that by using the inter-com I could talk, or rather sing, to my opposite number of the starboard side without it going to control below decks. All night it seemed we sang carols back and forth across the inter-com and the dark swell of the Atlantic.

Has ever a night of duffle dressed duty been more joyously spent? I doubt it. [282]

Christmas 1942 meant something very special to D.C. Laflin and David Riekie, both of Course 11, who were invited to the home of Mr. and Mrs. W.O. Fuller. *Courtesy Dr. and Mrs. J.E. Fuller*

American Families Take the Boys under Their Wings

The Daily Oklahoman visited Ponca City one month before Pearl Harbor and again, after the closing of the school, the author, Morris P. Moore, reported that pictures, often with gold stars placed on them, were found in multitudes of Ponca City living rooms along with stacks of letters from the young men who visited their homes.[283] Morris stated that Arthur E. Taft, an accountant with Continental Oil Company, and Mrs. Taft had entertained many boys during their time in Ponca City. He quoted Mr. Taft as saying:

Almost all the boys were delighted with America, its friendliness, its distances, its soda fountains and cars. Remarkably few were the least bit snobbish. They were usually shy and reticent at first but it was due more to a desire to avoid giving offense than to coldness. Upon cultivation, they proved warm-hearted, often affectionate, and surprisingly clean in thought and speech. They were intensely loyal and appreciative. Often they found it difficult to tell of their appreciation. They were not usually given to sentimental expression but never lacked the ability nor the will to write it. I think most of them would instantly accept the gage of battle from anyone who might slander Ponca City or Oklahoma.[284]

Another of the families who spent most of their time making the British lads feel at home was A.C. and Alta Dixon and their daughter Arduith Ann. One summer, Alta, whose husband was English, prepared many meals for the lads. She canned more than 700 quarts of fruit so that she would be able to serve them to her young guests. The Dixon family offered their home for other activities besides the sharing of meals. Often, the young men drew the blinds in her kitchen so that they could fill her kitchen sink with photo developer. Some of the lucky men were driven all over Oklahoma on short holidays by the Dixons even though the policy of rationing of rubber tires was in place. They took hundreds of "snaps" to send back to England of their adventures. Eventually, their families sent photos back to America so that the Dixons and others could see what home was like in England.[285]

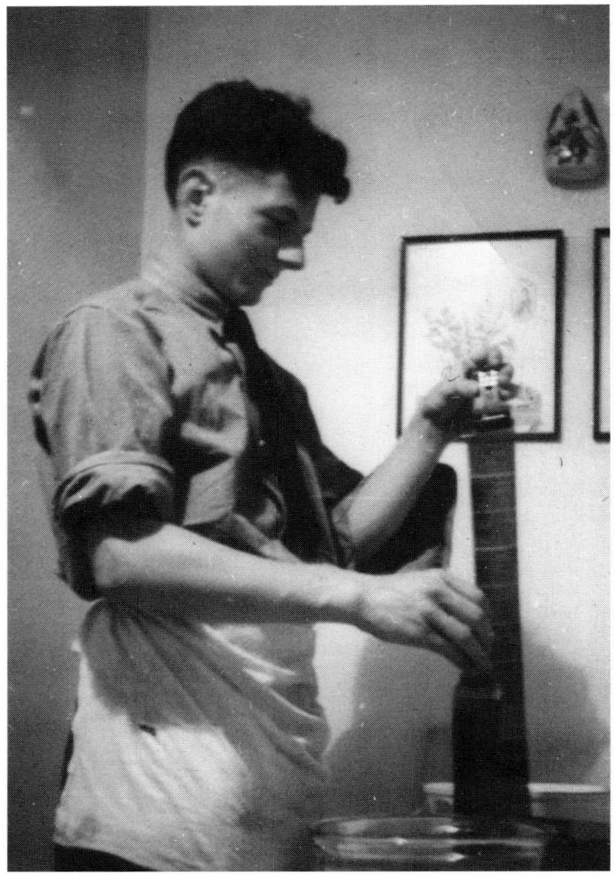

Jack Boswell, Course 9 at Darr School, was often found in the kitchen of Alta Dixon where he developed his photographs. *Courtesy John Cal Price*

Parishioners and RAF cadets visit after Sunday services held at the Episcopal Church in Ponca City. *Courtesy Dr. and Mrs. J.E. Fuller*

Two active citizens of the community, Mr. and Mrs. J.W. Dutton, served on various civic and church committees in Ponca City. Mrs. Dutton's granddaughter, Sarah Barrington Westerheide, described this time in the lives of her family:

Gladys Dutton was an excellent cook. She loved nothing better than to have friends into their home and to entertain. Thus began the routine of inviting several (whoever was free that night) "British boys" for dinner every Sunday and every Wednesday night. Their home at 340 South Palm was not a large home, but the dining room table always had room for one more chair. Wednesday dinner was "pot luck," but Sunday was always "waffle night." The young Brits had never had waffles before and they instantly became a favorite food. They would eat them with butter, with syrup, with honey, with jam or jelly and when the toppings were all gone, they would just eat them plain! Gladys would stay in the kitchen whipping up batch after batch until all had their fill.

Jim and Gladys, along with their grown daughters, Elsee Dee and Billie, became friends with numerous young pilots. It occurred to Gladys that the mothers of these young men might enjoy knowing how their sons were getting along. She said, "If my 18 or 19 year old son was far away overseas training to fight a war, I would want to know that he was safe." During the course of the evenings, after dinner, Gladys would get the names and addresses of the boys' parents and then sit down the next day and write them a letter. She told of meeting their sons, having them for dinner, and generally brought them up to date on their son's activities and well being. Two of the mothers to whom she became close were Ada Barrington, mother of Jack, Course 7, and Agnes Mary "Paddy" Igglesden, mother of Mike, Course 10. By the time Darr School closed in 1944, there were 86 British signatures in the Dutton's guestbook, each with an expression of gratitude for their hospitality.

Jim and Gladys Dutton's eldest daughter, Elsee "Dee," was employed at the Darr School as a control tower operator. At the time, she was one of the first female control tower operators in the United States. There was great surprise in the voices of the pilots responding to the female voice which came over the airwaves. Many aircraft to tower communications were inhibited at this point. Her job a Darr School gave Dee the opportunity to become friends with many of the young British pilots. One of their great delights was to "buzz" the control tower when they knew she was on duty. Their hope was to shake her up, but it never worked!

Mr. and Mrs. J.W. Dutton often invited their Ponca City friends and relatives to Sunday dinners where the special guests of honor were the British cadets. This picture shows a 1943 graduation dinner held for them in the Dutton home.
Courtesy Sarah Barrington Westerheide

LEFT: Harry Hill, of Course 11, says goodbye to Mrs. Alta Dixon as they wait for his train which was to return him to Moncton, New Brunswick before he was shipped back to Britain. Mrs. Dixon and her husband entertained dozens of young men while they trained in Ponca City. *Courtesy John Cal Price*

Westerheide went on to say:

One of the young pilots that Dee met was Jack Barrington. At the time, both she and Jack were engaged to other individuals, but they became good friends. Two of the interests they had in common were the love of dancing and of swimming. They would go as part of a group to various dances in Ponca City and became a very good dance team. Dee and Jack also enjoyed getting a group together on a regular basis for an afternoon of swimming at Wentz Pool.

Eventually, Dee accepted a new job and moved to Washington, D.C. She entered military intelligence service and subsequently joined the OSS (Office of Strategic Service). She was trained in ciphers and codes and was sent to London, England as a cryptographer. At this point, her mother, Gladys Dutton, wrote to some of her friends in England to let them know that Dee was on her way to London. These friends were eager to respond in kind; there was never a shortage of invitations for Dee to spend the weekend with her "English" families, especially the family of Jack Barrington.

ABOVE: Dee Dutton accompanies two of her English friends as they leave for England from the Ponca City train depot. Mike Igglesden is on the right but the cadet to the left is unknown. Later, Igglesden's father gave Dee away at her English marriage to Jack Barrington. *Courtesy Sarah Barrington Westerheide*

About her father, Sarah Barrington Westerheide wrote:

Born in Manchester, England, John S. "Jack" Barrington was one of the young British pilots that were sent to train at the Darr School in Ponca City, Oklahoma. He arrived in Ponca City in April of 1942 as part of 7 Course of No. 6 BFTS. He found the citizens caring and friendly. He especially enjoyed the invitations for "home cooked meals" and to the different functions held downtown.

After graduating the course in November of 1942, he became a flying instructor at an air base in south Texas, Kelly Field. He returned to Darr School as a flying instructor in August 1943. When No. 6 BFTS closed in May of 1944, Jack flew one of the last two AT-6's out of Ponca City, "buzzing" down Grand Avenue to the air base in Enid, Oklahoma. Upon his return to England in 1944, he was a Lancaster bomber pilot for the RAF for the remainder of European involvement in World War II. Shortly afterwards he was assigned to the Middle East.

Before heading for the Middle East, during one of his weekend passes, he was at home and learned that his mother, Ada, who was a "pen pal" of Gladys Dutton, had invited Dee for the weekend. The two friends met once again, finding that each was still single. Dee's fiancée was killed in action during the war and Jack's fiancée had literally written him a "Dear John" letter. They renewed their friendship and it quickly became a courtship. Although marriage was the goal, they both agreed that any wedding should take place after Jack had finished his service in the Middle East.

On December 20, 1947, Elsee Dee Dutton and John S. Barrington were married in London, England. Because her parents were not able to be at the ceremony, Dee was given in marriage by Hubert Igglesden, her "pseudo" father.

To this day the Barrington and Igglesden families remain close, the friendship having been continued to the third generation.

Jack and Dee Barrington returned to Ponca City in June of 1948. Both of their children were born and raised there – daughter Sarah Dee and son, John Christopher. Dee died in May 2, 1991 and Jack passed away in June 27, 1999. They are survived by their daughter and son-in-law, Sarah and Chuck Westerheide, son and daughter-in-law, Chris and Janel Barrington and three grandchildren – John Andrew Westerheide, Christopher Patrick Westerheide and Gretta Dee Barrington, all of whom share great pride in their parent's service to their countries.

Jack and Dee are laid to rest in Ponca City, where their story began.[286]

John S. "Jack" Barrington
Courtesy Sarah Barrington Westerheide

Arriving as one of the cadets in the first course at Ponca City was Scotsman Harold S. "Coggie" Cogman. He was one of the many young cadets who endured the hot Oklahoma sun while wearing a heavy grey wool suit and tie while exploring the town. Under order of the British Commonwealth, this was to be the accepted form of dress and the young men were very aware of the rules. He was one of four in his course who survived the war after flying more than 150 operations over Germany.

The jovial Scotsman usually had a joke or something to share with those who would listen. His fondest memories included the fact that many of the young men would often meet their dates at the fountain located in front of the city hall. After he retired, he was very active in his kitchen where he whipped up all kinds of breads and cookies.[287] Coggie returned to Ponca City and married Mary Sunshine McDaniel, a young woman from Tonkawa. The couple raised two children while he worked for Conoco Oil Company.

Elsie Dee Dutton poses in the home of Millie Van Winkle for her formal wedding portrait. She returned to England where she married Jack Barrington on December 20, 1947. *Courtesy Bret Carter*

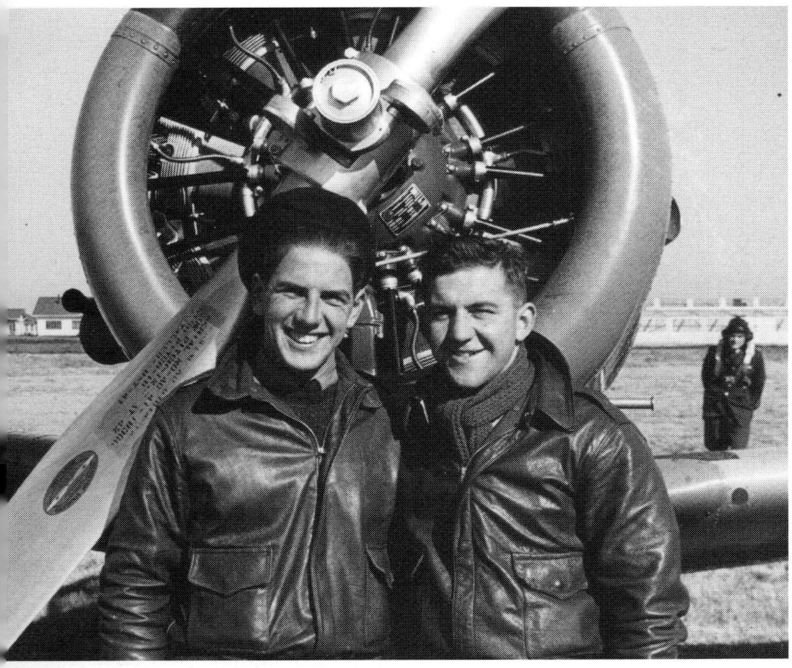

Harold Cogman, right, posed with his good friend, Tom Hennessy, who also survived the war. Many years later the two friends reunited when Hennessy returned to visit Coggie in Ponca City. *Courtesy Mrs. Harold Cogman*

In 1961, Coggie, who returned to Ponca City to live after the war, explained that while staying at the Ponca Military Academy, the fliers were overwhelmed with kindness. He described Mrs. William Cox, wife of the commander, who would make a daily trip to town to fill a variety of supplies, including paper and food–some too rich for the youth who were used to an austere diet. Coggie enjoyed the holidays in Ponca City. Christmas invitations were abundant and any given youth was invited to at least three or four dinners.[288]

When No. 6 BFTS Association began making plans for reunions[289] in Ponca City, he was helpful in encouraging former "fly-boys" to attend the reunions which occurred in 1984, 1989, and 1991.[290]

The Charles Hocker Family Adopts Many Cadets

Dr. and Mrs. Charles Hocker and their children, Jimmy and Iris, took many of the lonely young cadets into their Ponca City home as guests over a four-year period of time. Their residence, located at 1303 South 7th, was like a second home to many of the young men.[291] Always welcomed for scrumptious Sunday dinners, the hungry young men enjoyed tasting the wonderful T-bone steaks, cabbage salad, homemade ice cream, waffles, and fried chicken which was routinely offered. Eager to please and keep busy in their new surroundings, they were invited to visit and help at the Hocker farm located east of the city. When Dr. Hocker was not practicing dentistry, he was generally found working at the farm. Dr. Hocker owned a small farm and spent much of his spare time working there. Although it was definitely against RAF regulations, the young men took the chance and loved to "buzz" Dr. Hocker as they flew as low as they dared, just clearing the Hocker barn. What joy they had as they watched the doctor run out to search the sky and shake his fist, probably in jest.[292]

Mrs. Hocker helped on the farm and spent much of her time canning a variety of the fruits and vegetables grown there. She enjoyed knowing that the mothers of the young men living in England would be happy to know that the preserved food would be shared with the young men who visited her home. She and her family often took the young men to spend time in the country and they always remembered those times fondly.[293]

Jimmy and Iris Hocker stand in front of their Ponca City home at 1303 South 7th. The youth, children of Dr. and Mrs. Charles Hocker, became well-known and were enjoyed by the young men who came to know them as they visited as guests in their home. *Courtesy Mr. and Mrs. Jim Hocker*

As a variety of letters drifted back to Mrs. Hocker after the cadets' departures, she learned much about the families and customs of the men that she and her family came to love. It was evident that the men and their wives, moms, and widows were very fond of Verba Hocker. Each time they wrote to her, they considerately asked about the health of the doctor, who suffered from x-ray burns on his ankle, and about Jimmy's and Iris's school activities. Many offered advice to the children, stressing the importance of a good education. Some sent pictures of themselves, their new babies, and brides, but almost always the letters were addressed to Verba Hocker. She spent many hours answering these letters over a period of years. One of the brave young men, Jack Portway, wrote only once, with the letter marked later by Mrs. Hocker as "killed in action." [294]

No matter how hard the job of living day to day in battle or in homes lacking many of the vital foods and luxuries still enjoyed by the Americans, these British moms, wives, and the men always wrote with a positive and grateful frame of mind, thankful for what they had, although they had suffered through so much since 1939.[295]

One of the young men "adopted" by the Hocker family of Ponca City wrote a letter in which he gave a personal account of the flying bombs:

In June of this year Violet and I went to London for twelve days, about ten days after the flying bombs had started. After the first night when we had to dive under the bed because one of these "doodle bugs" engines had stopped just about overhead and came down very near to us, I became a widower for Violet would not sleep upstairs but only in the air raid shelter and I wasn't leaving my bed. It was quite a different experience to the bombing of London, because then there were hundreds of Barrage Balloons up and the guns would be firing for all they were worth – but with the flying bombs there were no balloons in London nor guns, and people used to switch their radios off as soon as the warning went so that they could hear if the flying bombs were coming near, because most of them when they were going to come down – their engines stopped. For five days of that leave we went to my home in Kent and there we were able to see them coming over and the fighters chasing them. Just in one afternoon we saw about twelve shot down, sometimes they burst in the air but in the majority of cases they came down and exploded on the ground. They used to come over quite low, about as high as we used to fly over your house in Ponca City. [296]

Years after the English had departed from Oklahoma, the old farm house owned by the Hocker family still stood. Many of the young men wrote to the Hocker family after their return to England and even years later, always remembering the times they cherished with the Hocker family at their home and farm. *Courtesy Mr. and Mrs. Jim Hocker*

Open Post Magazine had a warning for playful pilots upon their return to England in consideration of Barrage Balloons:

Do Be Kind to the Barrage Balloon

You have often seen them, floating above London or other large towns, looking as lazy and harmless a zoo elephants; but they have earned the right to appear languid because after all, they have accomplished their primary purpose: the prevention of dive bombing of military objectives.

You will soon be looking at them from a different angle and perhaps you should know a little more about them; for one thing the Balloon Barrage has been considerably improved since the outbreak of war and various "wicked" devices have been attached which give little chance of escape to the venturesome aircraft which blunders gaily into the cable.

There have been tales of Fighter Boys with an hour to spare, dodging between the cables and executing steep turns around the unhappy bladder but any such incident is always reported and becomes the subject of a visit to the "office" for the offending pilot. When on Ops you will always know the height is above ground level at this point and the Altimeter on the winch always under reads! So don't try to "just miss this one" as the one you didn't see half a mile off may be 200 feet higher.

Though the Balloon crew will not care about your breaking your neck, they will be most annoyed about the loss of their Kite. They will have to work like slaves to rig and inflate another and as the actual inflation always takes place at dawn or dusk any remarks made concerning you will be at least uncomplimentary. These balloons aren't festooning the sky for a purely decorative end. They are for a real purpose; so please restrain your playfulness. [297]

The wife of one cadet also wrote to the Hocker family:

Our little home had a bad knocking about but we are very thankful to be able to still live in it as most around us have lost their homes completely. The shock upset our nerves rather badly but we have been away for a while and are feeling very much better and it is very much quieter now and we are able to sleep at nights. It was the flying bombs that we found so terrifying. At one time we had no less than 150 over in 18 hours. [298]

Barrage Balloon *Courtesy Paula Denson*

Most of the cadets who returned to England were exposed to the bombs which wreaked havoc upon their homes or those which surrounded them. Stanley Nabbs, Course 14/15 described the flying bombs:

I expect you have read in the "Ponca City News" or some other daily paper about the flying bombs. Well, believe me, they are real swine! We have had three land around us, not one of which was more than 200 yards away. They have knocked out seven of our windows and broke the lock on the French windows by blast. Last week while we were at work one dropped 100 yards from us and brought the ceilings down on us. One of the boys had his head cut but mine being solid wood stood up to it O.K. [299]

Even as late as 1947, Mrs. Hocker was sending small but generous gifts and letters to the men and their families in England because the rationing and shortages of supplies did not stop for several years after the war. After receiving letters telling her that the children had never seen the electric lights turned on; that they had never seen bananas; that sugar was rationed to one half-pound per person per week; that only one ounce a week per person of cooking fat was allowed; and that at Christmas they were sometimes allowed about 1/10th extra portion of meat, one pound of oranges, one pound of nuts, and one egg,

Verba Hocker took action. She was compelled to share what she and her family had with her adopted boys. She often mailed Spam, rice, tins of baby food, cigarettes, cinnamon, pudding mixes, peaches, dates, nylon stockings, baby gifts, chocolate, and Spry, a shortening used for baking. In return, she received newsy letters telling her that Christmas cakes were made or that the babies looked adorable in the lovely frocks and dressing gowns she had sent. Each letter written by the mothers and wives and men was treasured by Mrs. Hocker and carefully tucked away in a shoe box, which she saved until her death in 1997.[300] Mrs. Hocker was preceded in death by her daughter, Iris Hocker Whitesell.

Ellen Bolton, one of the appreciative mothers, described her large garden of flowers and vegetables. She was very proud of her raspberries, which she had bottled as jam–also jellies of cherries, plums, and currants–and she was looking forward to gathering in her corn and picking hops in the early part of September, 1944.[301] Ellen's son, Dick Bolton, Course 9, and his wife, Violet, also wrote many letters to the Hocker family. They loved the Hocker family so much so that when their first child was born in December, 1945, they named the little girl Frances Verba, in honor of Mrs. Hocker.[302]

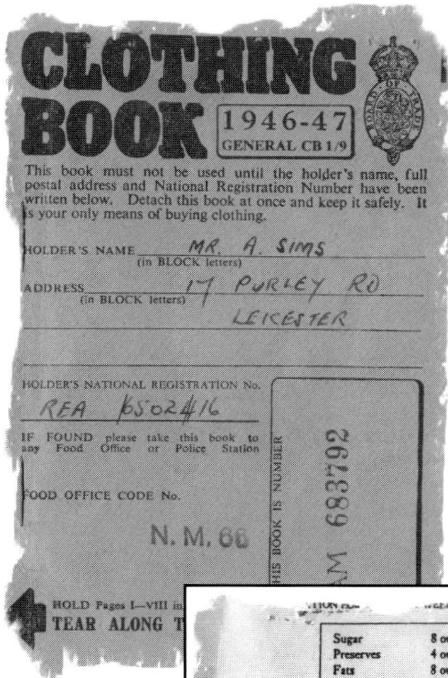

CLOTHING BOOK 1946-47 GENERAL CB 1/9

This book must not be used until the holder's name, full postal address and National Registration Number have been written below. Detach this book at once and keep it safely. It is your only means of buying clothing.

HOLDER'S NAME (in BLOCK letters) MR. A. SIMS

ADDRESS (in BLOCK letters) 17 PURLEY RD LEICESTER

HOLDER'S NATIONAL REGISTRATION No. REA 65024/16

IF FOUND please take this book to any Food Office or Police Station

FOOD OFFICE CODE No. N. M. 66

THIS BOOK IS NUMBER AM 683792

HOLD Pages I—VIII in
TEAR ALONG T

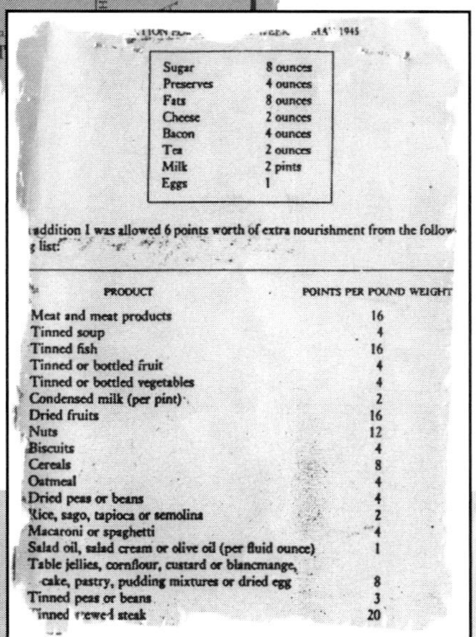

Sugar	8 ounces
Preserves	4 ounces
Fats	8 ounces
Cheese	2 ounces
Bacon	4 ounces
Tea	2 ounces
Milk	2 pints
Eggs	1

addition I was allowed 6 points worth of extra nourishment from the follow-g list:

PRODUCT	POINTS PER POUND WEIGHT
Meat and meat products	16
Tinned soup	4
Tinned fish	16
Tinned or bottled fruit	4
Tinned or bottled vegetables	4
Condensed milk (per pint)	2
Dried fruits	16
Nuts	12
Biscuits	4
Cereals	8
Oatmeal	4
Dried peas or beans	4
Rice, sago, tapioca or semolina	2
Macaroni or spaghetti	4
Salad oil, salad cream or olive oil (per fluid ounce)	1
Table jellies, cornflour, custard or blancmange, cake, pastry, pudding mixtures or dried egg	8
Tinned peas or beans	3
Tinned stewed steak	20

Clothing was carefully rationed during World War II and the years following, as seen in this 1946-1947 ration book. Mrs. Hocker often sent material goods, baby clothes, food, and other items needed by her English boys and their families. *Courtesy Albert Sims*

The Bombing of Hiroshima was the turning point in ending World War II. This photo shows one of the atomic bomb tests at Bikini Atoll as it was viewed by the late Hugh Wiles of Ponca City. He served on the USS *Flusser*, DD 368. Years after the war came to an end, food and supplies were still being rationed in Great Britain. *Courtesy Phyllis Mann Wiles*

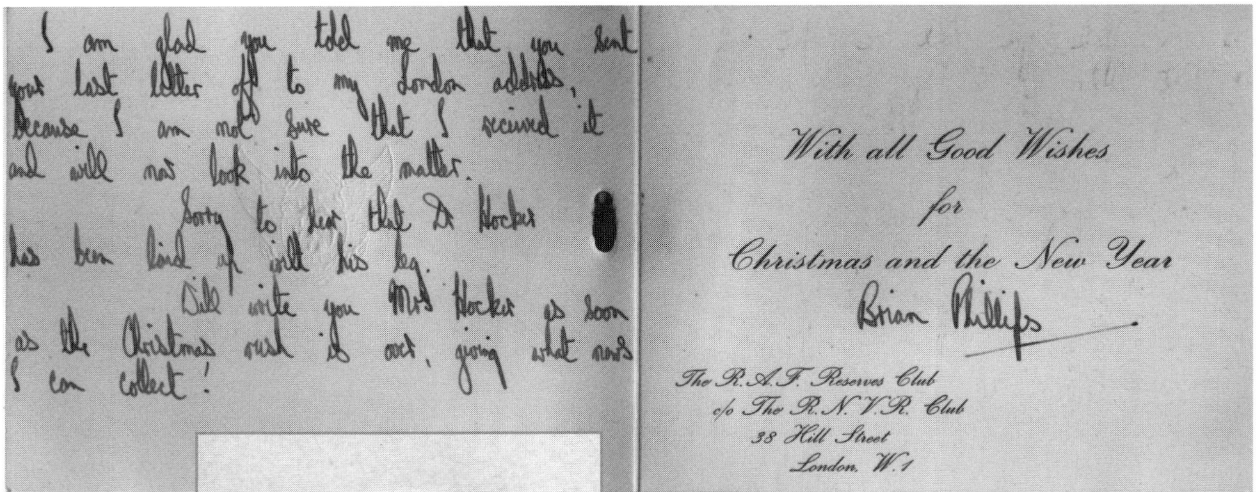

I am glad you told me that you sent your last letter off to my London address, because I am not sure that I received it and will now look into the matter.

Sorry to hear that Dr Hocker has been laid up with his leg.

Will write you Mrs Hocker as soon as the Christmas rush is over, giving what news I can collect!

With all Good Wishes
for
Christmas and the New Year
Brian Phillips

The R.A.F. Reserves Club
c/o The R.N.V.R. Club
38 Hill Street
London, W.1

ABOVE: This Christmas card was one of many sent to the Hocker family from their RAF friends after their departure from Oklahoma. *Courtesy Paula Denson*

LEFT: The RAF emblem, surrounded by an embossed border, was featured on the cover of one of the Christmas cards received by the Hocker family. *Courtesy Paula Denson*

Gladys Olorenshaw, mother of David Olorenshaw of Course 9 in Ponca City, moved with her husband, an employee for the Inland Revenue, from Chesham, near London, to Colwyn Bay, in Wales, during the war. She wrote to Mrs. Hocker quite often, always offering her thoughts about the state of the war and about the surroundings in which she lived. In one of her letters she wrote of her days spent on her uncle's farm as a young girl and shared her daily life by describing things such as her garden:

We don't bottle peas in this country. I wonder how you do it and if they are the same sort of peas? We don't bottle green beans either; I don't think we grow them here. The only vegetables we preserve are tomatoes which we bottle and green runner beans which we shred and pack in salt in jars. Actually, we have plenty of vegetables all the winter for on our allotment we grow kale, cabbage, cauliflower, broccoli, spinach, carrots and onions. Most of these come in for winter. We, of course, have potatoes and Brussels's Sprouts. In summer we have lots of peas; French beans and scarlet runner beans and lots of salads.

This year we are very short of fruit – almost the whole country had severe frosts in May which killed all the plums and most of the soft fruit such as raspberries and black and red currants. We do miss the fruit very badly. Back in April a good compliment of oranges and lemons arrived and didn't we all revel in them! We had pancakes with lemon, lemon drinks, lemon on all sorts of puddings and I made a nice bit of lemon curd. It did us all good for I believe we really needed it. I daresay we'll get some more before long but just now all the ships are needed to carry men and all the countless things they need to fight this great battle in Normandy. No one minds about such little things as lemons these days and all is going well, isn't it? It looks as if the Russians will be in East Prussia very soon now and we all hope the Germans will be beaten this year. Then there will be the Japs. Your men are doing splendidly against them! [303]

Gladys Olorenshaw's feelings for the Germans were plain–she felt disgust as she mentioned the particulars then coming to light about the awful concentration camps like Belsen. She said:

I would like to collect great numbers of the friends of Germany and make them go to see for themselves. For instance, de Valera; some Spanish friends of Hitler's and many people from all

Because England had to feed thousands of men who fought to defend their country, many foods were rationed to the general public and the people were encouraged to use every spare inch of ground to grow fresh vegetables and fruits. "Dig for Victory" banners were posted all over the country in an effort to encourage people to grow gardens. *Courtesy Imperial War Museum, London (D650A)*

neutral countries who think the Germans aren't really so bad. It is a terrifying thought how quickly a nation can degenerate into much beastiality. What you hear of these camps is almost unbelievable. There are lots of people even here who would do with an eye opener! [304]

Sometime during 1948, after her divorce from David's father, Gladys moved back to the London area where she resided with her parents. She later moved to Worthing, on the south coast and then to Rugby in the Midlands, where her sister lived. Eventually she moved to a smaller house in Syston, Leicestershire, where she died in 1979 at the age of 85.[305] In some of her letters to Dr. and Mrs. Hocker, she mentioned her two sons–David, who was awarded the Distinguished Flying Cross,[306] and Brian, who was injured in Germany and received a brand new medication to cure him. It was called penicillin.[307]

Mrs. Gladys Olorenshaw in front of her home at Colwyn Bay in Wales. She wrote many letters to her friend, Verba Hocker, who befriended her son during World War II. *Courtesy David and Birthe Olorenshaw*

David felt fortunate to have survived the war. He was flying Ansons after his return from Ponca City to England and at that time he was training gunners to shoot at a drogue pulled by another aircraft. He explained that:

One day, both engines went out so I had to land in the Channel - in the middle of winter. Luckily, the aircraft floated for a while and most managed to get into the half inflated dinghy with our unconscious instructor. I was busy sending Mayday signals and by the time I was ready to join the others, the dinghy had drifted a good bit away. My flying helmet prevented me from hearing the others call. I had to swim for it and they managed to haul me onboard, after which I passed out. Eventually we were picked up by a sand dredger and taken to a hospital, all OK, but looking like death warmed up because the yellow fluorescents from our life jackets stuck to our faces when we dove down to the bottom of the dinghy to find various tools.[308]

David was later posted to Bomber Command, flying Lancasters. He did 39 operations but suffered from appendicitis on the last one. He always had insisted that his crew members know at least one other operation on board so they could replace each other in case of injury or worse, death. The mechanic took over until they reached England, when they managed to get David into his seat to do the landing. He was rushed to the hospital where he had an appendectomy and recuperated with his parents at Colwyn Bay.[309]

David Olorenshaw and his wife, Birthe, visited with his adopted mother, Verba Hocker, center, when he returned to Ponca City many years after the war. *Courtesy David Olorenshaw*

David Olorenshaw was, of course, only one of thousands who had siblings fighting in the war and this must have caused pure agony for the families who waited for their return. Consider the parents of Peter Greaves, Course 10 at Darr School. He was the youngest of four children, three of whom were boys. Peter related:

Wilf and Don, being older, had both enlisted in the Territorial Army, which is the English equivalent to the army reserves. As soon as the war began in September 1939, the two older boys became involved, finding their way first to battle in Norway and then to Egypt, where they reinforced the English army in the Western Desert, in a major conflict that was about to begin and which would continue for three or four more years. They became part of the famous "Desert Rat," and played their part in such notorious engagements as Tobruk, El Almein, and Benghazi.[310]

They eventually came back home but by that time, Peter was learning to fly in Oklahoma.

After Mrs. Hocker's death in 1997 a family auction was held and a shabby shoe box full of the letters was offered for sale. It was then that Paula Denson purchased the box of treasured letters so carefully kept by Verba Hocker over the years. Surprised and intrigued by the contents discovered in the box, she began to seek the whereabouts of these long lost contacts with an earlier connection to Oklahoma history. Ultimately, after intensive research and much blind luck, some of the men were found to still be alive. One of them is John Cal Price of Course 17, who visited Mrs. Hocker upon his return to the 1991 reunion in Ponca City.[311]

English citizens used every spare inch of ground as they tended their Victory Gardens. This garden grows below a train line. Vegetables were vital to the survival of the English. *Courtesy Imperial War Museum, London (HU63756)*

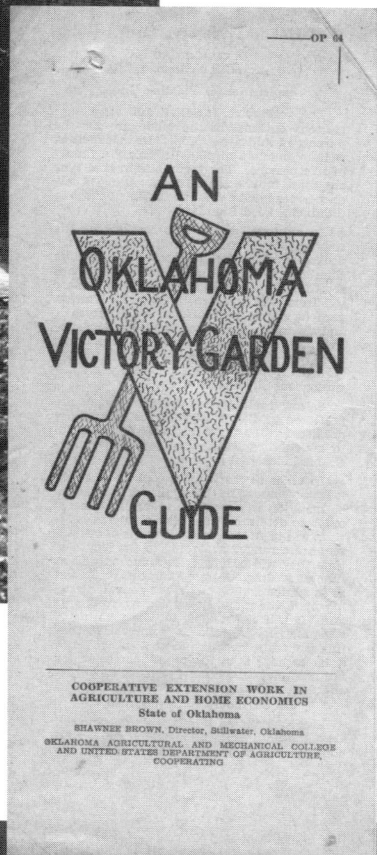

England was not the only country encouraging its citizens to grow gardens. An Oklahoma Victory Garden Guide was issued by the Oklahoma Cooperative Extension working in Agriculture and Home Economics. The brochure listed a variety of foods suitable to be cultivated in Oklahoma. *Courtesy Paula Denson*

BELOW: David Olorenshaw, top row, second left, poses with his Lancaster bomber crew. *Courtesy David Olorenshaw*

Not everyone who wrote the letters treasured by Mrs. Hocker for so many years were found, but their letters were so poignant that those who read them can understand that these people regarded their Oklahoma friends with much respect. Ella Brown and her husband, Norman Brown, of Course 15, sent many letters to Mrs. Hocker expressing their thanks for the gifts which were sent. Ella told Mrs. Hocker that they were still severely rationed for all food and clothes with little prospect of improvement that year. As civilians, they were not given as much as those in the military forces. They had been cut to one ounce of fat per week. In addition, they received three ounces of butter, three ounces of margarine and one ounce of lard. Because she was pregnant, Mrs. Brown was allowed an extra pint of milk a day and seven ounces of meat per week, along with about a pound per month of oranges and grapefruit. She was excited because the first allocation of bananas was to be given to the children of London for the first time in six years.[312]

Mrs. Brown wrote again to report that a shipment of tinned fruits, of the Libby's and Del Monte labels, was to arrive in the shops that month. She explained that because the fruit was worth 20 points per tin, she would not be able to purchase much, due to the fact that each person was allowed 28 points per month. These points could be used to buy tinned meats, fish, dried eggs, dried fruit, cereals, suet, jam and marmalade, jellies, and hosts of other things needed to prepare a meal. Mrs. Brown went on to report that they had a very plentiful plum crop in the summer and that she had bottled quite a few jars to store for winter puddings. She also had bottled tomatoes for her small son's breakfasts.[313]

Dr. and Mrs. Hocker were not always pleased by the information they received in their letters from England. A widow wrote to give the news about how she was dealing with the loss of her husband in early 1944. In the same letter she announced the birth of her newborn son. She described his weight and that he was very active, with gray-blue eyes and light brown hair. She had just learned that six of her husband's flight crew had been buried in a city cemetery near Munchen Gladbach, Germany but there was still no word about her husband, although she feared that he also had been killed at that time. She then went on to describe how her little girl looked very much like her father. She thanked the Hockers for the packets of puddings which they had sent.[314]

Another one of Mrs. Hocker's boys, Bob McFadyen, of Course 11, was killed in action, leaving a wife and an infant son whom he had never seen. One of her "adopted" sons replied to a letter in which she mentioned Bob's death and explained that he knew that Bob had been followed by other chaps who were trained with them and he felt particularly sad that Bob and his wife had so little time together. This flight officer offered the following information about the weekly war rationing in his 1945 Christmas letter:

2 oz. butter, one ounce cooking fat, 4 oz. margarine, 8 oz. sugar, 3 oz. bacon or ham, 1 egg if lucky, 2 oz. cheese, 3 oz. chocolate or sweets, 4 oz. jam or marmalade, 2.5 oz. tea, and about 1/2 pound of steak or two lamb chops.[315]

In his next letter he explained that coal and fuel also were rationed for heating and that there was grave danger of an extreme shortage that winter. Clothing was difficult for women to obtain when they were only issued fifty points for the year. For example, a frock required 12 points; a coat, 18 points; silk stocking, 3 points; and handkerchiefs, 2 points per coupon. Wives tended to "borrow" their husbands coupons to supplement their clothing.[316]

Many concerns voiced repeatedly in the letters sent to Ponca City related the indecision of the young men who had recently been "demobbed" and were struggling to decide what occupation to pursue. Many returned to the jobs they held prior to the war, because the British government ruled that these jobs should be held for them upon their return. Some struggled with the decision whether or not to invest many more years of their young lives going to school before entering the work force. A variety of jobs were mentioned in the letters including the selling of "tyres," insurance, and one man who was training for a supervisory position in the laundry business.

The other great concern voiced through the letters was the serious shortage of housing. The young men and their brides were living in the homes of their in-laws because one in four homes had been damaged or destroyed from bombing and available homes were selling for three to four times their pre-war value. Waiting lists were long. All this added to the other

Girls with automobiles were a big hit with the English, many of whom had never driven an automobile. It was against the rules to drive in America but the men soon found plenty of pretty young Oklahoma women who were willing to run them around town in "Daddy's" car. *Courtesy Mariann Bryant*

Young Phyllis Mann met her hero, Joe Moyle, and took him home for the whole family to enjoy. Although Moyle entertained her younger brother with his antics, he also found plenty of time to spend with his teenage girl friend. They kept in contact for many years as friends and Moyle visited Phyllis and her husband several times over the years. Moyle died in 2005. *Courtesy Phyllis Mann Wiles*

British Cadet Alan Watson proudly displays his newly-earned wings. He was one of those of Course 11 who graduated from No. 6 BFTS and returned to England as a flying instructor. *Courtesy Kristin Weltge Webb*

problems of day-to-day living faced by the young couples, who often had to wait several years before starting their families.[317]

Mrs. Hocker's heart broke over and over again as she read the sad stories which came her way. Perhaps she kept all the letters as a constant reminder of the privileges she continued to enjoy for so many years as she lived freely in a democracy bought and paid for by her English sons, as well as the Americans all around her.

Many years after the war, as surviving men returned to Ponca City in a succession of Darr School reunions held in 1984, 1989, and again in 1991, these "sons" returned to visit with Mrs. Hocker and proudly had their pictures taken next to one of the moms they loved so well. In their hearts, they all wished to tell the Hocker family the same thing that one of the young men who sent letters to the Hockers wrote:

I'll drink to you on the night peace is declared coupling with that silent toast–thanks which I have never adequately offered for the deep kindness you always showed me! [318]

Help, I'm Falling in Love!

The natural affection that some young men felt for the pretty girls in Oklahoma was certainly understandable and although studies were of the utmost priority to the cadets, their time spent in town with the ladies was anticipated with eagerness. Virile young cadets in Ponca City were warned by their superiors in subtle ways to keep their wits about them when it came to the young women they were dating. Nonetheless, some fell head over heels in love.

More than one of the handsome fliers had a difficult time leaving his "favorite girl" as he returned to England with the almost certain knowledge that they would never see one another again.

In March, 1985, Bill Bishop, a former RAF cadet who trained at Falcon Field in Mesa, Arizona, during World War II, wrote a letter to Charlotte Taft Weltge, living in Ponca City. During her high school years, Charlotte's parents, "Mom and Pop Taft," had hosted several young men from No. 6 BFTS, including Alan Watson of Course 11. Mr. Bishop explained in his letter to Charlotte that Johnny O'Brien, a Falcon Field, Arizona classmate, had later been assigned with Alan Watson to the same command at Scone, Perthshire, England, where the two men were flying instructors. During 1944 the two spent most of their off-duty time together and in September of 1944, Alan was attached to 78 Squadron Brighton for operational experience. He did two or three trips as one of a supernumerary crew on Halifax III aircraft and he went to "Happy Valley" in the Ruhr, after which he came back with a closer understanding of "flak." Prior to his attachment he wrote his first will and gave it to Johnny O'Brien for safe keeping.[319] Mr. Bishop said:

On December 5, 1944, while instructing Pilot Officer Wilson in aerobatics in a De Havilland Tiger Moth, the metal slot on the upper starboard wing broke loose and broke some wing struts in its fall. The upper wing collapsed and bent back covering P/O Wilson in the rear cockpit. The incident was witnessed by a Perthshire farmer in the village of Stanley who saw the pieces falling from the aircraft. He had the impression that the pilot in front was trying to shift the wing from the rear cockpit just before he (Alan) jumped. Having jumped at about 500 feet, Alan fell into the ground with arms outstretched and died at 3:20 p.m. He had not pulled the ripcord of his parachute, which was afterwards tested and found to be O.K.[320]

After Alan's death, his sister gave the young man's journal to Johnny O'Brien. According to Bill Bishop's letter to Charlotte, the sister reported that:

The fortnight's leave mentioned in the journal was disastrous from Alan's viewpoint. It was his first leave after returning from America and he was clearly full of all that had happened, his wings, his friendship with Charlotte and his close ties with the Taft family.[321]

Alan's mother, like many mothers, was clearly worried that Alan would return to Ponca City after the war and get married. She gave him a very rough time and ridiculed the feelings he had for Charlotte and her family. He was very hurt and upset and returned from his two weeks leave absolutely shattered. His journal ceased from that point and life was not the same.[322]

Since Mom and Pop Taft and their daughter, Charlotte Weltge, are deceased, no one will ever know the entire story. However, it is evident that Alan held a deep affection for the Taft family, who resided at 315 North 5th in Ponca City.[323] It is even more apparent that he was in love with Charlotte, although we do not know whether Charlotte felt the same towards Alan or if she only considered him to be a good friend. Charlotte wrote to Mr. Bishop, telling him that "one of the things I loved about Alan was his allegiance to country and family. Many of the walks and talks he speaks about were spent discussing our situation. When we said good bye at the train station we both knew it really was that. How he loved flying, family, and the heather on the hills near Hull."[324]

A saddened Charlotte Taft stands with Alan Watson, probably on the day he received his wings, since the cadets usually left Oklahoma the following day on their return to England. Alan clearly felt a great affection for Charlotte, as evidenced in his journal.

Courtesy Linda Colby and Tony Norton

These are excerpts from Alan Watson's personal journal detailing his time as he joins the RAF, his time at No. 6 BFTS and his love for a Ponca City girl.

Journal of RAF No. 1438732/150428
Alan Norton Watson
No. 6 British Flight Training School
Darr School of Aeronautics Student
Course 11, Ponca City, Oklahoma

Alan Norton Watson, RAF # 1438732/150428, volunteered for the RAF on April 15, 1941. He was born June 20, 1922 in Hull, England. His journal began on April 19, 1941 and ended on June 28, 1943. The following are only a few of the notations he made of this time in his life.

17th JUNE: *Reported at Jameson Street without gas mask and had to phone office to arrange for the civilian gas mask to be delivered and was then sent to Bedford on the train, changing at Leeds. At Bedford an RAF bus met the train and took us to Cardington camp, where there was a Barrage Balloon centre and, of course, the old Airship mooring mast. The next day I was given an exam to test all round knowledge and then a full aircrew medical. I was sworn into RAF on 19th June and given RAFVR badge (to wear in civilian clothes while waiting to be called up for training). Back home on the eve of 19th birthday.*

6th OCTOBER: *On the move on the train – but to where?*

7th OCTOBER: *Well, I have quite a lot to say in today's notes. We arrived this morning at Gourock on the Clyde at 5:30˙ and from the station we went to the dock and boarded the ship "King Edward." We waited on board this ship some time and at 1745 hours we left the shore and moved out to mid-Clyde. It was while we were moving in this direction that we noticed several aircraft carriers anchored, a heavy American cruiser with 2 Northrops aboard and about 12 merchant ships and numerous smaller ships, including invasion barges and RAF launches. Then I noticed a large 3 funneled liner some distance away and on asking a sailor what it was, I learned it was the "Queen Mary." That is the ship I am now aboard. We came alongside the QM and at first were rather disappointed at the size of her, but a more scrutinizing look told us the mistake we were making. We then transferred to this ship and went to our cabins. No 97, B Deck. I got to my bunk, one of the bottom ones and dumped my stuff and went for breakfast in the dining hall. The "Queen Mary" is certainly a magnificent ship. The food is very good, and everything else. At 6:15 p.m. we started moving, but it was only to the mouth of the river, where we stopped to take some more sailors and kitbags on board. There are a lot of American officers and men, with wives and kiddies on board. Am just going to bed now, the Lord knows what we shall be able to see when we wake up in the morning.*

8th OCTOBER: *This morning found us well under way with Ireland rapidly receding on our Port bow. The sea is getting pretty rough, the waves are already attaining a height of 30 to 40 feet and we are told that we are running – or sailing – with a gale and, believe me, I can quite believe it. The visibility is very poor today. The wind seems to be driving a mist before it, obscuring vision. I never imagined a ship of such tremendous dimensions as the "Queen Mary" could roll so much as she is doing. I've been up on the promenade deck nearly all today and I was amazed to see so few people on deck. It is indeed a job to walk up on deck today. There is a 50 mile an hour wind and that, coupled with the speed at which we are traveling, makes it impossible for you to keep your eyes open against the wind. This is certainly an epic adventure in my life. I never expected and I don't think anyone who has not seen the Atlantic can imagine how vast and powerfully magnetic the ocean is. I can quite understand the feelings of a sailor, loving the sea, but, despite the spirit of adventure the sea brings, I don't think it can compare in even a light degree with the air and flying. Here on the sea one feels so insignificant with the "Queen Mary" plowing on, the spray whipping over her bows and being flung back and driven over the forward part of the ship by the wind, which heaves through the rigging with an almost demonical fury. There is something fascinating about the rise and fall of the bows, as the "Queen Mary" forces her way ever onward despite the wind and the sea.*

I can only imagine what life is like on a small merchant ship on the Atlantic, but I can quite believe it is very hectic. All the RAF men on board are given definite duties to perform and Gordon (Knight) and I are supposed to be working in the Orderly room, but I am pleased to say that we choose our own hours. There are 18 of us in our cabin, fortunately I have got a bottom bunk. Therefore the air around me is the clearest in the cabin, even so, it's going to be pretty hot, I figure.

Things are getting rougher tonight. I am about to retire now for my second night aboard the "Queen Mary." I've just been to the canteen and bought some more oranges and chocolates and cigarettes. We can get as many of these as we wish. – Goodnight.

12th OCTOBER: *Went up on deck before breakfast and was amazed at the colour of the sea. It was a beautiful turquoise blue, or maybe slightly deeper and the sun, which was shining at that hour, made the whole scene really marvelous. The day,* *unfortunately, was spent in the Purser's office working and this evening in a bath of cold sea water reading. Now feeling more comfortable I am going to retire. Today we have been watching flying fish skimming over the ocean. They are wizard little things and fly for quite a distance before popping merrily back into the sea.*

(Note: I have not mentioned this before but a few days before we sailed the "Queen Mary" was involved in a collision with a cruiser and stove her bows in. This was temporarily repaired with the aid of concrete and she is going into dry dock in Boston, in the United States, for complete repairs. You can see the bent and jagged plates as you look over the side when right up forward, out she is running quite well despite this. Incidentally, the cruiser went down in a few minutes, the "Queen Mary" rode over her and smashed her back. Quite a lot of the crew was lost. The "Queen Mary" did not stop. She left the convoy which the cruiser was escorting to pick up survivors. "The ship will, under any circumstances, carry on – remember, a man overboard is a man lost" is the notice on the board).

14th OCTOBER: *What a day. I think this has been one of the most memorable days of the whole journey. At 10:45 this morning we saw our first glimpse of land for a week out this was lost to sight as a heavy sea mist fell. No one can quite interpret our feelings as all the men just looked at the land, bathed in sunshine, it was the dawn of an entirely new an different career, the start of a new series of experiences and adventure. At 4:20 p.m. we were pushed and maneuvered into opposition in the dock by eleven fussy little tugs and by 5:00 p.m. we were all tied up and secure. I was on deck all this time and could not help but remark on the shapes and sizes of the super cars that abound in the States. Contrary to our expectations we docked at Boston and not New York. Later in the evening I could only just stand and stare at the glistening lights reflected in the water and the flashing of Neon signs and powerful headlights of the cars as they sped around the town. It was a magnificent sight, not having seen land for seven days and then to see it as you haven't seen land for three years. At 8:30 p.m. we left the "Queen Mary" and boarded a train and then were on our way on a 24 hour journey across the States and up to Canada. I'm writing this on the train. We have opened the seats which come down to be beds and am just going to sleep. The "Queen Mary" as we crew say on the train, was an unforgettable sight bathed in light–truly marvelous.*

15th OCTOBER: *We are going to No. 6 BFTS Ponca City, Oklahoma.*

20th OCTOBER: *I've had a pretty lousy day today on the whole. First thing this morning I was transferred from 6 BFTS to 3 BFTS , Miami, Oklahoma. I was to leave all the boys and go with 3 BFTS and as I don't know any of the men going there, I did not want to be transferred. I went to the CO and all he could do was, if I could find someone in 6 BFTS who would rather go to 3, we could swap. I've spent all the day looking for someone and when I was just about to give up hope, I found a guy by the name of Cameron who wanted to change. We went to the CO and he said it would be OK.*

10TH NOVEMBER: *Today we left Moncton on our way to Ponca City. The route – Moncton – Quebec, Montreal, where we all went around, The Wax Museum, Notre Dame Cathedral, St. Lawrence River, two hectic hours, then two more hectic hours in Chicago. At Port Madison we crossed the Mississippi River. Kansas City where they again spent some time.*

13th NOVEMBER: *Arrived in Ponca City. Food terrific, including bananas.*

29th NOVEMBER: *Went into Ponca City today and had a look round. It's a very clean town and reminds me somewhat of an English seaside resort without the sea. Had lunch in the town and then went for a walk with Gordon and Reg. At 3:30 we went to a social at the First Baptist Church to which all our flight was invited. There we met three girls – Tommy, Virginia and Kathryn. I went for Tommy. She is a swell girl, 21 and incidentally has her own car.*

19th DECEMBER: *35 minutes dual today. It was very misty and we got lost. However, we found the 'drome and landed and stayed down. Went into town and met Mr. Taft who invited us to dinner and we had a grand time. Went to the Club Lido.*

FRIDAY, CHRISTMAS DAY: *Gordon and I got up at 10:30 and had breakfast and then we all went to see some friends of the Tafts and Gordon, Charlotte (Taft) and I stayed in the small park near their home and played like a trio of kids. We were very happy. It's Christmas and a time to have fun.*

3rd JANUARY: *I made three flights today and boy, do we do some flying nowadays. I spent the rest of the day at 315 with the Tafts. Gee, they are swell people.*

18th JANUARY: *Having lectures in the morning and I got the idea that as this bad weather seemed to be here to stay for some little time, we may as well have our leave this week. I suggested having leave this week, the Wing Commander agreed. We were paid, given our passes and just like that – we were on leave. I rang up Mrs. Taft, who very kindly invited me to spend the week with them and as I can't afford to go away this time, I really appreciate this. I guess I can't express my feelings in words. It's really great of them – they are marvelous. I went into Ponca to* *their house and in the evening I went with Charlotte to a show at the Poncan. Came back to 315 at about midnight. We went to the Grand Café and had to wait for a long time to be served.*

20th JANUARY: *Got up early in the morning and got a taxi out to the camp in order to pick up some clean clothes. In the evening Charlotte and I went to see "A Night to Remember." It was a good flic and I thoroughly enjoyed it. Charlotte is a swell kid, a great sport, and I like her an awful lot.*

22nd JANUARY: *I went with Mrs. Taft to the Continental and had lunch with Mr. Taft and Charlotte. It was very nice indeed. I'm certainly having a grand time. Mr. and Mrs. Taft have been wonderful. Words can never express my gratitude to them. I shall never forget them and their hospitality. In the evening we all went to the pictures to see "Sky Devils" with Spencer Tracy. I've seen it twice before but enjoyed it again nevertheless.*

27th JANUARY: *I flew this morning for the first time on A.T.6.A's or, as we call them, Harvards. They are lovely ships, altogether different from a Stearman. I'm going to like them a lot. In the evening I went into the town, took my battledress to be cleaned and spent the evening at Charlotte's.*

28th FEBRUARY: *I spent the day with Charlotte and Mr. and Mrs. Taft and we all went out to the lake and had a picnic. I took Charlotte to the flics in the evening.*

12th MARCH: *Charlotte came out in the car to see me this evening but I could not stay long as I was flying at 8:00 p.m. It was pretty grim too. (Flying, I mean).*

1st MAY: *Ground school in the morning and first swim in newly filled swimming pool at camp. Then I went into town and Charlotte and I went shopping for some things to take home for everyone. I got Tony a pair of genuine Indian moccasins and a few other trinkets. Also some underwear and stuff for Mum and Jerry.*

11th MAY: *The end of the wings exam.*

15th MAY: *Junior and I had a lot of fun this morning. He wanted to take some photographs of me in flight so we went out together, took the pics and then proceeded to shoot each other up. We were doing slow rolls at 200 feet. Grand fun. We played tag in the clouds. Sammy taxied into the crash truck. Went to the flics with Charlotte and stayed the night at their house.*

18th MAY: *Still as wet as hell. There are floods all over Oklahoma. Got the results of the Wings exam today and I was amazed to hear that I have to sit Meteorology again – the last subject I expected to fail as it's my strongest point. Charlotte and I went for a walk this evening – yes, in all the rain, and when we were out it started to pour. We walked on and then watched the lightning – which was really magnificent. Then we went to Sam's and played the Juke Box. "Rose Ann of Charing Cross" was our immediate favorite. I'm afraid it was only tonight that I realized how much I love Charlotte and how much I am going to miss her when I leave here. That tune will always be ours and will always remind me of this night in Sam's. I'll never forget it.*

21st MAY: *Had a final practice for our wings parade and then went out with Charlotte for a walk. Learned that I had passed my Met re-sit which I took this morning, so I sent a cable off to Mum and Dad. Feels good to know that get my wings tomorrow.*

22nd MAY: *Well, at last we have had the great day – the day for which I have been working for the past eighteen months-Wings Presentation. Mom, Pop and Charlotte were there and did I feel proud as Brigadier General Smith of the United States Army Air Corps pinned on my wings. Gee, it feels good to have these wings up.*

In the afternoon I could not go home with Mom and Pop and Charlotte, as I had to fly some more, so I went up with Ben Kelson. He made me fly the ship from the rear pit and I gave him a beautiful 35 minutes. It was the last flying I shall do before I get back to England. To round off a perfect day, Charlotte and I went to the flics together and had a wizard evening.

24th MAY: *I spent most of the morning filling in forms and things in reference to our commissions. Did a spot of shopping in the town in the afternoon and then on to 315 and then on to the wings dinner. It was a super affair. Addresses by the Wing Commander, "Casey" Jones, Captain Phillips and Bob Aston, our Cadet Wing Commander. It was an excellent show, held in the Conoco Cafeteria. Then Pop and I went home for Charlotte, after having quite a long chat with Mrs. Ball, who also comes from Hull. Then came the dance – my last evening in Ponca City with Charlotte. It was the most wonderful evening of my whole life. Charlotte and I danced every dance together – it was marvelous. Then we walked home with Dick Cleaver and Mildred (his girl) and stopped at a café and kept it open, whilst we ate and played "Rose Ann."*

24th MAY: *What a hell of a day! I am now on the train heading for Moncton. I've never felt so bad saying goodbye to anyone as I did when I left Charlotte and Mom and Pop. God, I wish I could take her with me. I'm going to miss her horribly. We had our last shoe-shine in Ponca and the train left at about 5:00 p.m. I'm not sure – I was in a daze. I'll never forget the look on Charlotte's face as I said goodbye. Her eyes were red and the wonderful kid was doing her best not to cry.*

28th MAY UNTIL JUNE 5th: *We saw a few films and then had no money and did nothing all day. Moncton is the lousiest hole I've ever been to in my life. I write to Charlotte every day and she writes me. I wish she were here.*

13th JUNE: *Next week I shall be 21. I went for a walk in the rain with Junior to find a juke box with "Rose Ann of Charring Cross" on it and was lucky enough to do so. It reminded me of Charlotte so much – not that I really needed the juke box.*

23rd JUNE: *We had breakfast at 6:30 a.m. and caught the train at 7:15. We arrived at Halifax to learn that we are going back on the "Louis Pasteur," a French liner. We boarded the ship and learned that I am in section E 10 with 120 other officers. There are bags of planes flying around. It reminds me of Ponca and Charlotte.*

24th JUNE: *At 2:00 p.m. tugs pulled us out from the quay and we started our return journey. It was with very mixed feelings that I watched the land slide beneath the horizon. Whilst glad I'm going home, I hate to leave the United States and Charlotte so far behind.*

25th JUNE: *I walked up and down the deck to pass the time and cannot help thinking of Charlotte.*

28th JUNE: *I paced the decks with Ken Grover for ages and then we had a boat drill and all the guns were fired. I was on deck and saw a wizard display. After dinner I listened to the gramophone some more, thinking of Charlotte.*

30th JUNE: *Ten days after my birthday we first sighted land to starboard, which was Northern Ireland. This was at 9:45 a.m. A little later we saw Scotland on the port bow and passed on down the Irish Sea. Further on we passed the Isle of Man to Starboard. Later we docked in Liverpool. Next day we left the ship and went to Harrogate, where we stayed waiting for leave for two weeks. Then two weeks leave, about which I need say nothing. And so our American adventure comes to an end.*[325]

Bill Bishop was able to share this journal with Charlotte Taft Weltge forty-one years after Alan Watson left Ponca City. In a short letter thanking him for the information, Mrs. Weltge, who was busy caring for her gravely-ill husband, mentioned that both she and Alan were aware of the fact that when he returned to England they would never see each other again.[326] Although each of the men in Ponca City and Miami had their own story, this one was quite typical of the feelings many of the men had as they left behind their girlfriends and adopted families to return to their homes in England and the war that lay ahead.

Leonard Joseph "Little Joe" Timperley is pictured on the right with John Hathaway. Both served in Course 13 at Darr School and on the day of their graduation, they were assisted in sewing on their wings by their friend, Ethmer Hansen, who was deft with a needle and thread. The men are pictured in Ethmer's garden. Timperley, who was twenty years old, died October 28, 1944 and is buried at the Chester (Bacon) Cemetery in England. *Courtesy John Hathaway*

Training Accidents Claimed Lives in Oklahoma and England

Flight Sergeant Leonard Joseph Timperley came to Ponca City as a 19-year-old Lancashire lad in March, 1943.[327] His story, although poignant, was not unique and could have easily been that of any of the brave young men who took their place in the fight against Hitler's Germany.

The son of Joseph and Elizabeth Timperley of Stretford, Lancashire, young Joe was eager to learn to fly and after his training in Ponca City, he returned to England where he became part of a crew of Wellington X HF465.[328] Not much is known about the lad, but it is believed that he was a happy-go-lucky young man who carried one of those tiny collapsible doll toys of a Hawaiian hula dancer on the dashboard of his plane.[329] While he trained at Darr, he was befriended by Ethmer Hansen, a Ponca City woman. She and her family often invited "Little Joe" and his friend, John Hathaway, to their home where they enjoyed "horsing around" in her garden. Timperley was awarded his wings on September 29, 1943. With Mrs. Hansen's help, she sewed the brand new wings and his sergeant's stripes onto his dress uniform before his return to Moncton on October 2.[330]

In August 1944 Flight Sergeant Timperley met the rest of his crew at No. 12 Operational Training Unit, Chipping Warden, Oxfordshire. Crewed-up in their final weeks of training before being sent to an operational station, Timperley was joined by bomb-aimer Sergeant Laurence Edwin "Laurie" Harrison, age 24; Pilot Officer Cecil Frank Holmes, 29, a navigator/bomber; Sergeant James David Milne, 32, a wireless operator/air gunner; Sergeant Arthur Fred Baker, 35, an air gunner; Sergeant Ronald Whiteley, 23, an air gunner; and Sergeant Douglas Stewart Hollis Wilson, 22, a navigator.[331]

On October 28, 1944, this crew, accompanied by an instructor, began a cross-country practice flight. Three hours and ten minutes into the exercise, a severe thunderstorm was encountered. The in-experienced pilot, trained with too few hours to fly a fabric covered war plane, suddenly experienced volatile weather and iced-up wings and could not handle the situation. The Wellington bomber began to break up and losing both wings, disintegrated on impact in a farmer's field.[332] Only one eight-year-old farm boy, with his uncle by his side, witnessed that horrible scene.[333] The navigator, thrown clear, landed with the aid of a parachute while the other six airmen and an instructor were killed.[334]

Bomb-aimer, Laurie Harrison, killed on the ill-fated Wellington, married his sweetheart, Marjorie, just four months earlier in a beautiful wedding ceremony. Marjorie Shelton, the widow, described her wedding to the handsome young RAF officer more than sixty years after his death:

Dressing up gave these ceremonies the importance they deserved and made them very special. Marriage was forever – a joyful yet solemn occasion in a church, with a wedding dress showing the proper respect for that institution. Up until the 30s, gloves were worn and a veil covered the face, but during the war, customs were already changing. I chose to forego the gloves, which had to be removed and given to the chief bridesmaid before the wedding ring was placed on one's finger, but instead had a pointed sleeve which reached down to the base of the fingers. I did keep the traditional wreath of orange blossoms around my head. One of my two aunts, who married before their men were sent abroad to fight, chose to wear a street dress–dress, coat, hat, gloves and a handbag with her prized fox fur piece and a flower corsage.[335]

Courtesy Marjorie Shelton

Marjorie Shelton was not the only one to walk down a church aisle only to return a few months later to repeat her steps as she attended her husband's funeral. Few realize that more than fifty percent of the Allied Air Force crew killed in World War II died on training missions such as this one and few memorials are erected to remember those who gave their lives to the war, although not in battle.[336] These sacrifices were no less than those who received glory for being shot down over foreign lands. Of course, many survived to return home to their brides. Pictured are several examples of RAF wedding photos shared with Ponca City friends.

Courtesy Dr. and Mrs. J.E. Fuller and Paula Denson

Harrison's widow, Marjorie, who remarried in 1948 and immigrated to Canada, returned to that farmer's field in England years later as a widow for the second time. Her search for the location at Prince Hill, Bridgemere, Cheshire, was not an easy one. Marjorie and her daughters began searching for clues and eventually found the "gentle hillside." In September, 2000, at the age of 77, accompanied by her daughters, Marjorie visited England again and at the exact spot where the fuselage of Laurie's plane came to rest, she placed white roses in the field in memory of the young men who died there.[337]

Two years later, in October, 2002, Marjorie was contacted by Derek Inskeep, a Cheshire historian who was interested in erecting a memorial near the site. The widow and her new friend from England began a search for information on the rest of the men involved in the tragic crash. As facts fell into place, Derek obtained scaled plans for a memorial and submitted a planning application to the Crewe & Nantwich Borough Council for their approval. Bridgemere Parish Council and a local monumental mason joined the project and a dedication ceremony was held on the closest Sunday to the 59th anniversary of the crash–October 26, 2003. As Marjorie and her daughters returned for this emotional service, they placed a bouquet of Rosemary for remembrance,Lavender for memories and peace, Holly for renewal and resurrection, Oak for England, and Laurel for the sacrifice of these men and all those who laid down their lives, whether in training or battle, no matter the land in which they fell.[338]

In Ponca City, seven young English cadets, three American students, and several instructors died while serving their countries. Perhaps the poem, penned by

Marjorie Shelton's daughter, Carol Jane Shelton, expresses the feelings of those left behind who still remember these deaths, no matter the location:

Go To Your Sweet Rest

It is a farmer's field, like any other
Where the wind blows free
And the songbirds gather.

There are spirits here, among the flowers,
Who walk the hedgerows and mark the hours
Seven brave souls whose wings were broken
But of their sacrifice no word is spoken.
It was up to those who were left behind
To bear the wounds with no reason or rhyme.
To make sense of a thing that made no sense at all,
Forfeiting lives full of promise
With those who answered the call.

It is just a farmer's field, like any other,
But here the wind can fall silent
And the songbirds will linger.
For spirits walk here, among the hallowed flowers,
Along the sacred hedgerows, spending timeless hours.
Seven young souls who went boldly forward,
Bearing old men's burdens, bravely shouldered.
It was up to those who begged the angels nightly,
To rebuild lives now shattered and empty.
Was it for glory of war or the will of God Almighty,
This dreadful price to pay for the love of country?

It will remain just a farmer's field, like any other,
Where the wind carries voices
And the songbirds answer.
For the spirits here, we lay wreaths and flowers,
By the hedgerow chapel, for a few brief hours.
To the seven precious souls who left the world too quickly,
We honor your sacrifice, your courage, your bravery.
You died not in vain but for some higher purpose so true
Now your memory lives on in hearts you never knew.
Go now, you young cavaliers of the sky,
Go to your sweet rest as we bid you goodbye. [339]

For the most part, spending time in Oklahoma signified some of the happiest times in the lives of the men who trained there. However, not all experiences for Darr School students in Ponca City were enjoyable. When one of the cadets contracted polio, Course 12 was quarantined to the base for their course's graduation and thereafter it was known as the "Polio Class." Fortunately, no others in the school developed the dreaded disease.[340]

The harsh reality of irrevocable loss set in quickly as several students and instructors died during flight training exercises. *The Ponca City News* issued the following bulletin:

An airplane believed to be from the Darr School of Aeronautics and piloted by a British student, crashed approximately two miles south and three miles west of Newkirk shortly before 3 o'clock Tuesday afternoon. First reports said one of the two men aboard was dead, the other severely hurt.

The next day another article reported the grim details. Instructor Henry Dalrymple was killed, along with British flying student, Raymond Haslam. The plane was virtually broken in two by the impact.[341] Later, the newspaper also recorded the funeral:

Townspeople, flight instructors and British student fliers filled the Grace Episcopal Church Thursday to pay tribute to Raymond Haslam, 21, Manchester, England, who was killed in an airplane crash Tuesday near Newkirk. An eighteen man guard of honor composed of British cadets formed in front of the church after the service.

A uniformed detachment of American Legionnaires attended and accompanied the cortege to the I.O.O.F. Cemetery, where the youth was buried.

While the Reverend Gordon Smith intoned the graveside rites, airplanes piloted by British students and their American instructors droned in the distance. Then the British guard of honor marched individually to the grave, came to attention, and gave the farewell salute. After saluting, the officers and men about-faced and formed west of the grave. The British flag which draped the casket was removed. Casket bearers were British students, Kenneth Charles Weller, Cyril Philip Bailie, Herbert Ibbotson, Glyn Hankins, Gerald Watson and John Barry Noble.

Flowers from commanding officers, instructors, employees and every department at the Darr School of Aeronautics, as well as from many townspeople, were tributes. The E.M. Trout Funeral Home was in charge of the arrangements. [342]

Not long after the funeral, an expression of thanks was sent to Ponca City citizens:

Crash Victim's Parents Express Gratitude for Ponca Citians' Kindness

Mr. and Mrs. E. Haslam, Manchester, England, parents of Raymond Haslam, British flying cadet killed near Ponca City January 27 in an airplane crash, expressed their gratitude to Poncans for their sympathy and service.

The following note was sent by the Haslam family:

"Mr. and Mrs. E. Haslam, Lucie, John and Dorothy desire to express their sincere thanks for the love and sympathy shown by all at Ponca City to their dearly loved son and brother, Raymond, who lost his life in an aircraft accident while training for his country's service. They also wish to thank Rev. Gordon Smith, the county attorney, Roy Grantham, and all who sent flowers in caring sympathy.

He gave his life while training for his country in a righteous cause for the survival of justice and freedom for all mankind.[343]

This was the beginning of the loss of several British students who trained in Ponca City.

On September 30, 1942, a mid-air collision occurred between two students and their instructors. One of the instructors and his student were able to bail out after a power dive from the other plane went through the tail assembly of their trainer at 2,300 feet. The survivors met on the ground and set off in different directions to seek assistance. Eventually the cadet met a small boy who guided him to what he thought was the wreckage of his own aircraft. Sadly, it was the aircraft which had hit his tail and both the occupants were still in their seats, beyond human help. Those killed were student J.A. Neild, 19, and instructor Marvin Hill, 32.[344]

One of the aircraft crashes experienced by students in Ponca City. It is not known which specific incident is pictured, but this reflects a typical crash site scene as authorities approached the accidents. *Courtesy Ponca City Library*

Seven English graves stand side by side in the Ponca City I.O.O.F. Cemetery. This photo was taken before all of the men had died. Each year they are honored with special ceremonies held on Memorial Day. *Courtesy Phyllis Mann Wiles*

In another fatal crash, *The Ponca City News* also reported on January 7, 1944:

A British cadet and his instructor from the Darr School of Aeronautics at the Ponca City Airport were reported killed when their training plane crashed in the Osage hills six miles northwest of Pawhuska, at 3 a.m. Friday, according to the Associated Press.

A cowboy is reported to have spotted the plane's wreckage and to have telephoned the information to Pawhuska. Police Chief L.O. Maze and a squad of Pawhuska auxiliary highway patrolmen reached the plane shortly before noon and reported by radio that both occupants were dead.

Two squads of auxiliary police and a two-way radio ambulance were rushed to the scene from Pawhuska, and an army jeep with crash crew, a station wagon containing officers, and an army ambulance were reported sent from Ponca City.

The names of the instructor and cadet were to be released by authorities at the Darr School when next of kin had been identified. [345]

The student was later identified as number 1807591 Harry A. Dinnage, of Course 15 and the instructor was A.A. Neumann.[346]

Cadet Dinnage's death was a bitter experience for his mother far away in England. When one of the chaps who knew Harry returned to his homeland, he took the time to visit with the cadet's mother. She had received Harry's personal effects but was very unhappy with the Royal Air Force because they had taken income tax out of his pay. Although the young pilot officer who visited her promised to return for further visits, he was so bothered by her sadness that he never followed through.[347]

No. 6 British Flying Training School
requests the pleasure of your company
at a Graduation Dinner
at the Continental Associates Building, seven o'clock
Wednesday, 1st of December 1943
followed by a dance at the Club Lido

R. S. V. P. Dancing 9:00 - 2:00 Semi-formal

Most courses issued engraved invitations to Ponca City friends for the dinners and ceremonies which were held for the new graduates of Darr School. *Courtesy John Cal Price*

Even though a few students and instructors died in training in Ponca City, Jack Barrington related that the skill and dedication of the Darr instructors was so great that "the RAF could later boast that some of its finest pilots were graduated from this civilian-contract program, of which Ponca City ranked among the best. As an indication of the professionalism of these American instructors, Ponca City had the lowest accident rate of all American flying training schools." [348]

In all, there are seven graves of cadets located side by side in the Ponca City I.O.O.F. Cemetery. They are: [349]

• 1497498	RAYMOND HASLAM	Course 5	1/27/42
• 1512082	JOSEPH A. NEILD	Course 9	9/30/42
• 1317871	JOHN F. SWAIN	Course 7	11/5/42
• 1653024	OWEN W. PHILLIPS	Course 14	7/5/43
• 1580817	GERALD O.G. WORRALL	Course 13	8/19/43
• 1807591	HARRY A. DINNAGE	Course 15	1/7/44
• 1806046	R.D.W. (JUMBO) EYRES	Course 17	1/18/44

Graduation

As each course graduated from No. 6 BFTS, they were honored with a Wings Parade, at which time they were presented with their wings by their commanding officers. As most courses graduated, Conoco invited the cadets to "Wings Dinners and Dances" hosted by Continental Oil Company and their junior associates.[350] It was not uncommon for several hundred people to attend the celebration and ceremonies, including the students, staff, and American friends. In April, 1942 the No. 6 BFTS graduation class was served a meal of roast beef and Yorkshire pudding in the spacious cafeteria of Continental Oil Company.

Some of the British youth were often invited into private homes for special celebrations after their graduations and the men, eager to have their wings sewn on to their uniforms, were quickly helped as adopted moms and girlfriends pulled out their sewing baskets to help attach the wings. There was always both laughter and crying at these happy occasions.

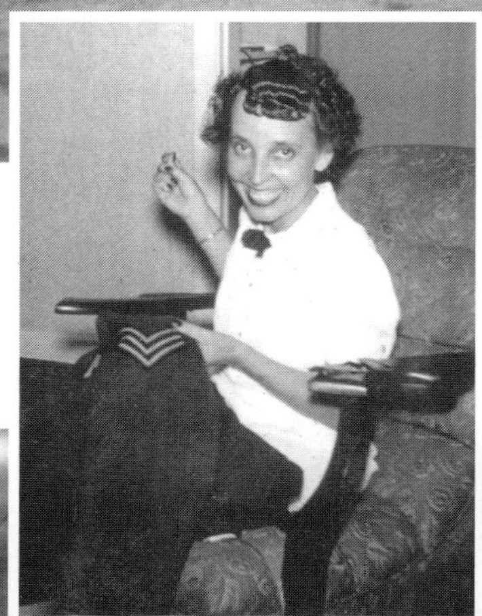

Billie Smyer happily sews on the wings of one of her friends as he graduated from No. 6 BFTS. *Courtesy Mariann Bryant*

ABOVE: Following the presentation of wings to Darr School graduates, the Wings Parade passes by a review stand of RAF and American officials. Lined up in the background is a large group of AT6 Texan aircraft. *Courtesy Dr. and Mrs. J.E. Fuller*

Course 12 at Darr School was known as the "polio" class because one of the cadets developed the dreaded disease while in Ponca City. The entire class was quarantined to the base and held their graduation dinner and party in their barracks. Despite the fact there were no women present, they managed to make the best of a bad situation. *Courtesy Ponca City Library*

Darr School of Aeronautics Closes

The United States War Department took over Darr's contract with the British on April 1, 1942 while the Defense Plant Corporation (DPC) bought the real property and leased it back to Hal Darr and his school. The DPC called the site PLANCOR 1482.[351] In April, 1944 the United States Army Air Force began to slow the production of its own pilots drastically.[352] At the same time, Wing Commander A.V. Rogers announced that the Royal Air Force delegation in Washington, D.C. had ordered that Darr School of Aeronautics would be closed and pilot instructors and other United States Army personnel were ordered to leave by April 15.[353] Security officers were asked to remain for a period of time until the government could decide what should be done with the buildings. Sixty mechanics were left to prepare the aircraft for transfer to other fields.[354] Members of Course 17, who had not completed their training at the school, were dispersed for the completion of their graduation at several of the other schools including Miami, Oklahoma, Mesa, Arizona, Terrell, Texas and Clewiston, Florida.[355]

A conductor announces their pending departure as English cadets, sporting newly-presented wings, prepare to leave Ponca City on the train which will return them to Moncton, New Brunswick before departure back to England and the war. Crowds of Ponca City residents always gave the young men a special send off. *Courtesy Mariann Bryant and John Cal Price*

Mike Igglesden, left, was a member of Course 10. He is shown leaning out the window of his departing train accompanied by his unidentified friend. More than sixty years later, Mike has been instrumental in the successful planning of many of the No. 6 BFTS Association reunions held in Great Britain. *Courtesy Sarah Barrington Westerheide*

Many years after the closing of the school in April, 1944, one of the barracks of Darr School stands alone, a visible reminder of No. 6 BFTS. *Courtesy Bret Carter*

At the closing of the Ponca City school in May, 1944, the ensign was presented by the last commander to Colonel Charles Bradlaugh Thompson, who faithfully drove from his home in Tulsa each Decoration Day to fly the British flag over the seven graves of the young fliers buried in Ponca City. Thompson, earlier in his life, came from Lancaster, England and traveled to the United States where he found a job in Nevada working for an Indian boarding school. While he was there he met and married a daughter of Ar-ke-ke-tah, the last hereditary chief of the Otoes. He and his family later moved to Oklahoma.[356] At the 1982 reunion of No. 3 BFTS in Miami, he traveled to Miami from his Tulsa nursing home for that occasion. This time he came as Sir Charles Thompson. In 1969 he was presented as a Member of the Order of the British Empire (MBE) by Queen Elizabeth for his work at the Ponca City school where he served the school and tended graves.[357] He is buried along side the seven RAF cadets who died in flight training accidents where they are interred at the I.O.O.F. Cemetery in Ponca City.

Immediately after the close of the school, the Ponca City Chamber of Commerce conducted a planning meeting in the Jens-Marie Hotel grill room where they considered possibilities for the continued use of the Darr School buildings. Government agencies would not allow commercial businesses to use the school facilities and so the Chamber of Commerce decided to prepare a prospectus which would be submitted to the federal board of hospitalization and the veteran's administration, in the hope that the buildings might be used for a veterans hospital or rehabilitation center.[358] No action was ever taken by the federal government and by fall of 1944 more than 500 aircraft were stored as surplus at the defunct school.[359] Row upon row of aircraft were lined up across the field as seven mechanics preserved them until they could be put up for sale using a bidding process at a sales center located on Cimarron Field in Oklahoma City.[360] Eventually the remaining aircraft were disposed of and, by means of an interim permit for use of the property in 1946, Piper Aircraft established a factory, although it closed early in 1948.[361] Most of the vacant buildings were gradually disassembled and in July, 1948, the City of Ponca City was given permanent possession of the property from the war assets administration.[362]

In October, 1941, several years before the actual closing of the school, Squadron leader Suren's comments truly reflected the feelings of those who remained in the last course to train in Ponca City. He said:

It is safe to say, when the 20 week turn over hits its stride, many British boys will carry with them over Germany and the occupied countries visions remote from the scene below. They'll see not pick-marked objectives, but the rolling green of Kay County wheat, Lake Ponca sparkling below, the Pioneer Woman's bronze glow in the bronze autumn landscape, an Indian in braids, the Conoco tank farm and perhaps the dancing eyes of a Ponca high school girl. For Ponca has learned to love the British boys, and the boys love Ponca City. [363]

A few days after Darr School was closed, an anonymous writer wrote the following:

They came and they conquered the hearts and friendship of all who came to know them, those British gentlemen who as aviation cadets learned to fly here and who have just departed for homeland to battle bravely the enemy.

Words seem not to sufficiently express the admiration for the conduct of these young men – polite, cheerful, polished, well-spoken and truly representing the backbone of British youth. To see these RAF boys with their American cadet pals in training, getting along so well together made the heart feel warm for this American-British friendship and gave bold assurance to the truth that American and British people do get along well together despite Nazi sympathy and propaganda which occasionally attempts adverse criticism.

We shall miss those gentlemen, even at night when the skies are still and quiet. We shall feel lonely but comforted in knowing that you also shall miss us in Ponca City. Good luck and God speed you on to victory. [364]

Years after the war was over, most other men who trained at the two Oklahoma schools were busy leading diverse and fulfilling lives as they contributed to the world around them through their work and generosity. One of them, Reg Gilbert, Course 14 at Darr School, helped establish a trust in 1988 which helps young people traveling in unfamiliar lands to get a close-up feel for the country and an understanding of its culture and people.[365] Each year the Reg Gilbert International Youth Friendship Trust (GIFT) distributes small bursaries to youth, ages 14 to 25, who wish to visit a developing country such as Guyana, Kenya, Jamaica, Chile, The Gambia, Nepal, Israel, Mexico, Sulawesi, India, Poland, Uganda, Thailand, Sri Lanka, Tanzania, Ecuador, and Jordan. For the duration of their visits to these countries, the youth spend time as an ordinary member of a typical family and often work in schools and hospitals or help in other community projects.[366] Reg Gilbert was awarded the Order of the British Empire, where he was presented to Queen Elizabeth in recognition for his world-wide contribution to humanity.[367]

A Bittersweet Return

As each course of young men from No. 6 BFTS and No. 3 BFTS returned to England, it was with the somber realization that many of them would not live to see the hard-fought freedom they desperately sought. However, these brave young men knew that without these sacrifices, the hope of peace for the world that they and their families would live in was a "pipe dream." There are countless stories of those young men who lost their lives. The following is another typical example of the loss of life for some of the young men trained in Oklahoma:

Flying Officer George Such, of Course 5, was awarded his pilot's wings and commission on his 21st birthday, August 7, 1942 in Ponca City, Oklahoma.[368]

In 2004, his brother, Arthur Such, of Redditch, England, contacted the author asking for information about his late brother, George G. Such. At that time, his only picture of his brother was one showing him attired in civilian clothes. His brother was searching for help locating a photograph of George in his RAF uniform. This picture was discovered in the photo album of the late Alta Dixon and was personally presented to him in October of the same year. George, left, enjoyed horseback riding with his friend George Major. *Courtesy John Cal Price*

While he trained in Ponca City, Mr. C.R. Galloway, his advanced flying instructor, often commented that Such was the smartest pilot he had ever seen.[369] After further training on heavy bombers after his return to England, he and his crew commenced operations over Germany on the May 13, 1943 with 76 Squadron from Linton-on-Ouse, Yorkshire, later moving to Holme on Spalding Moor, Yorkshire, in June, 1943.[370]

The squadron completed thirteen operations, including one aborted mission due to engine failure. Their fourteenth and final operation was over Hamburg on the night of Saturday, July 24, 1943.[371]

Flying Officer Such had carried out two operations as a 2nd pilot, which was standard procedure for all bomber pilots prior to commencing operations with their own crew. One of these operations was over Cologne where Such was 2nd pilot to the commanding officer of the station.[372]

Their aircraft, a Halifax DK 187M, was shot down by a German night fighter at 12:10 a.m. on the morning of July 25 off the coast of Holstein at the dog leg turn where the pathfinders had dropped flares to indicate the turning point. Above them, the night fighter picked up DK 187M because of the light from the flare. Many crews reported in their debriefing upon return to base that they saw the Halifax fall in flames and explode before reaching sea level some fifty miles out from the German North Sea coast as it turned toward Hamburg.[373]

The body of Sgt. Jackson, 2nd pilot on his first mission, was washed ashore ten days later. Flying Officer Such remained on the missing-in-action list until 1949 when his body was discovered by the British War Graves Commission while looking for missing RAF aircrews. He was buried in a church cemetery in Holstein, Germany. The six remaining members of his crew were never found. Their names are recorded on the Royal Air Force Memorial, Runnymede, England, along with the names of 20,000 other RAF air crew members who lost their lives in World War II and who have no known grave.[374]

In eleven weeks, 76 Squadron lost 60 percent of their aircraft. A tour of duty required a crew to carry out thirty operations before being rested. With the average crew strength of twenty aircraft being available at any one time, 76 Squadron lost 59 aircraft for the year 1943.[375]

Flying Officer George Such received medals including the 1939/1945 Star, Aircrew Europe Star, Defense Medal and War Medal 1939/1945.[376]

The men of No. 3 BFTS and No. 6 BFTS returned to England to finish what the Germans had started in 1939. It would be impossible to relate even a very small amount of the information gleaned from the stories of each of the approximately 3,500 men from both schools but a common theme developed in respect to their war-time duties. Many from the earliest graduating courses served as pilots on coastal commands, night fighter bombers, flew PR Spitfires, served in the Middle East or Cairo, worked in India, or photographed as much of China as they could. Others flew an immense variety of aircraft during their careers.

Wing Commander Jerry Eaton, of Course 6 at Darr School, was head of H.Q. Maintenance Command at RAF Andover before his retirement in 1970. During his career, in addition to the training planes flown in the United States, he flew the following aircraft: Miles Master Jks 1,2, & 3; Mustang 1 & 4; Tiger Moth; Spitfire Mks 4,9,11,13,16; Hurricane Mk 1; Proctor 3; Typhoon Mk 1A; Tempest Mk 5; Provost Mk 1 (piston); Jet Provost T4; Meteor Mks 7,8,11,12, & 14; Vampire Mks 3,5, & T11; Venom Mks 2 & 3; Prentice; Magister; Hunter; Javelin Mks 2,5,7,9, & T3; and a Power Chute Kestrel (powered parachute). Together with an old friend who also was an ex-Typhoon pilot, the retired wing commander was invited to Normandy for several ceremonies honoring those Typhoon pilots who were killed in action on or about D Day. As part of his duties, Eaton unveiled a plaque set up by villagers of Caen, who wanted to honor the memory of one of the pilots of No. 257 Burma Squadron who was shot down by flak near that village. Between May and August, 1944, 151 pilots lost their lives in that area.[377]

Jack Youens flew Mosquitoes, Ansons and Defiants and was part of 21 Squadron and also part of 140 Wing which included 464 Australian and 487 New Zealand Squadrons of 2nd tactical Air Force. Today he maintains active communication with many of the men who served with him.[378]

Life after the War

Many men never returned home, finding a place of peaceful repose on the hillsides of England or in military cemeteries scattered throughout Europe and the Far East. Those who did return were eager to get back to the business of living. It was not easy for them. There was a drastic shortage of housing due to the bombings of the war and because many useful commodities, such as building materials, food, and clothing, were exported. It took years before many of the men felt that they could take on the responsibilities of marriage and a home.[379] The English government provided for the men to return to their previous civil positions if they chose, but many were not happy with that arrangement. Some loved flying so much they stayed in the military, while others flew as commercial airline pilots. Noel Clark and David Olorenshaw are only two of those who chose this route.[380] Others returned to school and became teachers or businessmen. A host became jewelers, botanists, insurance agents, cab drivers, bankers, and laundry men. Some joined the new age of computers as the industry progressed. Almost all fell into lives of usefulness and service to others.

Letters written by Alf Salkin, Course 14 at Darr School, were discovered in 1997 among those written to the Hocker family in Ponca City. However, it took more than 60 years after he lived in Ponca City and nearly eight years of research stretching all the way to Australia before anything was learned of the fate of the young man.

Alf Salkin was posted to Operation Training Unit Number 71 in Cairo, Egypt and later in Italy, pictured here as a warrant officer. He flew Spitfires during this period of service. After the war, Alf took art lessons in Italy and eventually migrated to Australia where he earned a Bachelor of Arts and a Master of Science at Monash University. He was well known for his avid involvement in the maintenance of the natural bush lands of Waverly, Australia. *Courtesy Ian Salkin*

Alf Salkin, kneeling, on top of the Baw Baw with his Australian friends and his brother, center, dressed in black. *Courtesy Mr. and Mrs. James Hocker*

According to Ian and Owen Salkin, Alf spent his early childhood living around Manchester, England. From a family living in a "precarious financial position," he moved with his family from place to place. While the family lived with a friend of his mother whom they called "Aunt Ada," the woman introduced him to walking in the Penines, a low range of hills between Lancashire and Yorkshire. This was the beginning of his love for nature. After several years, the young Jewish boy and his family moved to Higher Broughton. While attending school, Alf was introduced to the planting of small bulbs and quickly learned that he enjoyed seeing the bulb grow into something beautiful. Again, his love of nature was stirred.

To help pay the rent, Alf would accompany his uncles to the local town markets where he worked all day for the sum of fifteen cents. As he grew older, he learned to repair bicycles. When the war broke out, he began a course to become a fitter and learned rather quickly, being accustomed to the use of a file. Eventually, he worked at Trafford Park as a fitter for the Shell Petroleum Company maintenance depot and his work was considered to be one of reserved occupation status. However, he soon changed jobs and began working on magnetos.

Alf and his family moved again after the home they lived in was bombed. The family was soon asked if they would billet airmen who were to be posted overseas and Alf quickly learned that most of them were not much older than he. They were going to various locations with the Empire Air Training Scheme in order to learn to fly. Alf's interest in the possibility of joining the RAF as a flyer grew and in 1942 he applied for training. After he was accepted he began a series of studies including the Theory of Flight, Engines, and Navigation. Eventually, he was sent to Heaton Park and soon began his trip on the *Queen Mary* as she made her way across the North Atlantic. Alf quickly learned that Prime Minister Winston Churchill was aboard. After a short stay in Canada, he was assigned to Darr School of Aeronautics in Ponca City, Oklahoma.

Upon Alf's return to England following graduation as a Flight Sergeant in Ponca City, he was posted to a holding unit at Harrogate where he waited to go on operations. While training in an advanced flying unit Alf met a number of Australians. Another friend from New Zealand described the mountains as being very much like those of the Himalayas because of the glaciations and Alf was fascinated.

After this training course, Alf was sent in a troopship through the straits of Gibraltar, traveling then through the Mediterranean until they reached Alexandria, Egypt. Along the way he met many more Australians and again, he found them interesting.

While he was stationed near the Suez Canal he flew his first Spitfire VC. He continued flying them in the Middle East and in Italy, later converting to flying Mustangs as a part of the 112th RAF Sharks Squadron. When the war ended and while Alf was still in Italy with the RAF, he studied art. Returning to Manchester, Alf worked in the clothing trade with his brother, Ivor. Soon, both men knew that they wanted to migrate to New Zealand. However, they had no luck accomplishing this goal and after a short time they learned that Australia needed machinists. Changing plans and only three weeks later, they were on their way, arriving in June of 1949. They immediately began working in their trade as machinists but, because the work was dictated by the seasons, they found they had plenty of time to ride bikes cross country or to hitchhike as they explored the Australian countryside. Their home base was in Melbourne. Eventually, both men trained as teachers. Alf's interest in art caused him to become an art teacher. Soon he began to teach at Balwyn High School and Mount Waverly High School. On December 12, 1959 he married his wife, Esma, and they made their home in Mount Waverly, where they raised two sons. As a family, they went on many adventures, walking in the bush and exploring the wonders of nature. Eventually he taught at other schools including Brentwood High School, working as the principal of the middle school until his retirement.

His love of music, reading, art, and the environment were always apparent to those around him. With Esma, he became heavily involved in maintaining the natural bush lands around Waverly and was honored in the Monash World Environment Day Awards as the winner of the Community Leadership Award in 2004.

Throughout his twenty years of retirement he and his wife traveled the world. On March 7, 2005, Alf passed away.

Return to Oklahoma

Several former RAF cadets came back to Oklahoma to continue their lives after the war. Three of the men who trained in Ponca City eventually returned there to live and raise their families. Harold "Coggie" Cogman[381] and Albert "Bert" Sims[382] both became employees of Conoco while Jack Barrington joined an insurance firm.[383] Each married a Ponca City girl and the three families became active in their church and community events. During World War II and shortly after, "more than two dozen American-trained British fliers married American women whom they met during their training."[384]

As the years rolled by, small groups of the men joined together to form associations and since their initial conception, a series of reunions have been held world-wide so that former friends and instructors may reunite in order to talk of those days so long ago. The group from No. 6 BFTS met in 1984, 1989, 1991, and 1994 in Ponca City. They still meet annually in England. At the No. 6 BFTS reunion held at Mickleover, England in 2004, Air Vice-Marshal H. Gill shared one of his many observations of his early days in the Royal Air Force during World War II.

If you came down in the ocean; if you could get out of your aircraft; if your rubber dinghy inflated; and if you could get in it, your salvation rested upon your ability to blow the whistle so hard that a ship one hundred to one thousand miles away would hear it and come to your rescue! [385]

Albert "Bert" Sims poses as a new officer of the RAF.
Courtesy Albert Sims

Albert Sims, back row center, smiles with his Lancaster crew during a break between bombing missions over Germany. All survived the war and although they are scattered throughout the world, they have kept in touch by "Round Robin" letters ever since. *Courtesy Albert Sims*

American cadet, John Wessale, left, with his RAF buddy, Tony Wadley, both members of No. 3 BFTS. *Courtesy Tony Wadley*

Cadets from No. 3 British Flying Training School wait their turn to learn new flying maneuvers shortly after their arrival in Miami, Oklahoma. *Courtesy Deryk Maddox*

CHAPTER THREE
Number Three British Flying Training School Spartan School of Aeronautics Miami, Oklahoma

On July 13, 1941, and only three weeks before the Darr School of Aeronautics commenced their operation in Ponca City, No. 3 British Flying Training School opened its doors for training of English pilots in Miami, Oklahoma, a small town located on historic Highway Route 66 in the northeast corner of the state.[386] Its establishment made the third Spartan Aeronautics Company School dedicated for the training of pilots in Oklahoma. The former two schools were established for the purpose of training civilians and located in Muskogee and Tulsa.[387] Some of the new cadets assigned to the flying school in Miami were especially excited when they learned of their destination until they discovered that the palm trees and sandy beaches they were expecting were located in Florida–not Oklahoma.[388] Miami's roster included 2,124 RAF cadets and 117 USAAF cadets of which 1,493 pilots were awarded wings upon graduation.[389]

How Miami was Chosen

The Miami Chamber of Commerce first submitted plans for the site of an air school in 1940 when Mr. Harry Berkey, secretary to the Miami Chamber of Commerce, joined with other members of the chamber in presenting the idea that a training school for RAF pilots should be located there.[390] According to the *Miami Daily News-Record* issue of October 17, 1982, Harry Berkey reported that he traveled to St. Louis, Missouri, to visit with the Clayton Knight Committee. There he proposed the idea of a British Flying Training School in Miami. The Clayton Knight Committee was developed in England to recruit American pilots to ferry aircraft for them to England. Because the committee's role was to recruit pilots and not to train them, they suggested that he carry his plan to the British Embassy. Returning to Oklahoma, Berkey and his committee, including W.J. Martin, chief engineer of the NEO Railroad, developed "A Proposal for the Number One Royal Air Force Pilot Training Facility in the United States." Berkey was accompanied by Herb B. Cobban and Pop Root when they presented the proposal to the British Embassy in Washington, D.C. in the spring of 1941. Lord Halifax told the committee that if the Lend-Lease Bill was passed he would consider Miami for a school.

After passage of the Lend-Lease Act, Captain Maxwell Balfour, CEO of Spartan School of Aeronautics in Tulsa and J. Paul Getty, president of the company, were invited along with other American contract pilot training schools, to travel to Washington, D.C. There they were encouraged to consider establishing Royal Air Force training schools. Balfour told this meeting that he liked the idea but that there was no room for a school at their facility in Tulsa. In response, the British Embassy gave Balfour the information provided by the Miami Chamber of Commerce. Later, because Balfour visited Miami several times, the city's hopes were high that they would be chosen. However, Balfour chose Ponca City instead.[391]

When Maxwell W. Balfour became director of Spartan on July 1, 1939, his previous experiences completely qualified him for his new role at the school in Tulsa. During World War I he had driven an ambulance in France. Then he learned to fly in 1917 and was schooled in several French locations before becoming a member of the Third Pursuit Group.

The Spartan School of Aeronautics sign welcomed new cadets who arrived for each of the 25 courses graduated at the new school in Miami, Oklahoma. *Courtesy Ponca City Library*

Newly-arriving cadets searched the streets of Miami for pretty girls upon their arrival at No. 3 BFTS. This cartoon was drawn by one of the cadets who attended the school. *Courtesy Ponca City Library*

Several years later, Balfour served at the American Embassy in Paris before returning to the United States where he served as adjutant of the Fifth Observation Squadron at Michell Field. Men who flew aircraft which crashed and lived to tell about it became members of the "Caterpillar Club." Balfour joined the club in 1928 when the wings were torn off his plane during a roll. He was only 600 feet above the ground when he was able to jump from the plane. A year later, another of his planes caught on fire. Captain Balfour waited to jump while trying to coax his passenger to jump from the burning plane. Unfortunately, they waited too long and the plane crashed, killing the passenger and burning Captain Balfour severely. He spent three years in the hospital while he recovered. Upon his release from the hospital, he was retired from active military duty. At that time he had more than 5,000 flying hours.[392]

Early one morning soon after Balfour's decision to locate in Ponca City, Mr. Berkey and Mr. Herb Cobban, president of the North Eastern Oklahoma Railroad, paid Balfour a call in Tulsa. Cobban is credited with changing Captain Balfour's mind. So convincing was his argument to locate the school in Miami that Captain Balfour exclaimed, "For two cents I will set up the school in Miami instead of Ponca City!" Herb Cobban pulled two pennies from his trouser pocket and the tryst was sealed.[393] Thus, No. 3 British Flying Training School was born.

The buildings at Miami's school were constructed in a design very similar to structures used in Ponca City. However, the administration building contained three stories and housed a knotty pine paneled PX and lounge.[394] A spacious, three-story building, formerly named the Pierce Pennant Terminal Building, had been purchased for use as the executive offices for the new British Flying Training School. Originally built as a hotel, it was constructed in 1929 but had lain vacant for five years preceding the opening of the school. The ten acres of land and the building had been owned by Sinclair Oil Company and were located north of Miami on Highway 66.[395] Squadron Leader A.C. Kermode related that the site selected near the administration building was a large, waving cornfield before building commenced.[396] The Miami Chamber of Commerce purchased 80 acres of land west of the G.A.R. Cemetery along with the Pierce-Pennant Building.[397]

Captain Maxwell Balfour, CEO of Spartan School of Aeronautics, leaves his aircraft as he visits the new school at Miami for final inspection. *Courtesy Ponca City Library*

ABOVE: Not all cadets were able to travel to homes such as the one of "Mama" Jenks. This group of Miami trainees decided to make the best of their situation and play with "Re-Pete", a German Shepherd adopted by the entire school.
Courtesy Northeastern Oklahoma A&M College

LEFT: An aerial view of Spartan School of Aeronautics in Miami, Oklahoma looking west toward the distant hangars.
Courtesy Ottawa County Historical Society

After Miami's selection as a Royal Air Force training school, plans were immediately established by Maxwell Balfour to begin the construction of barracks which would house the 200 British chaps expected to arrive within weeks. Two mammoth hangars also were included in construction plans as well as five other buildings. In order to pour enough concrete for the huge hangars, two railroad car-loads of sand were used each day during their construction.[398] By July 17 the newspaper reported that the Waller-Wells Construction Company, under the supervision of A.L. Ward, had completed a cafeteria and that classrooms and barracks would be finished within eight days. Other projects in various phases of construction were a flight building for officers, a 37,000 square-foot hangar including aprons, and a 4,000-foot-long by 100-foot-wide all-weather runway, built by the Gaines Brothers Construction Company on 80 acres.[399] The Muskogee Iron Works and the Allied Steel Firm of Tulsa were responsible for the construction of the hangar. Three other auxiliary fields also were constructed shortly thereafter.[400] The J.E. Mattern Construction Company and Hogle Plumbing were used as contractors during the building of the school along with the Oklahoma Electric Company of Muskogee.[401] Monetary outlay expected as the school's construction began totaled about $250,000. Ultimately, Spartan would invest roughly $500,000 in the establishment of the school.[402]

Captain Balfour named Robeson P. Tucker of Tulsa as one of the flight commanders, along with Jack Casey and Anthony Ming. Other advanced training flight commanders were H.E. Bollar and S.D. Bassett.[403] As the school geared up to begin, Balfour announced that twenty-five flight instructors were to report in the next week or so and that they would begin training in eighteen primary training planes. The Vultee BT13A aircraft were to be flown in from Randolph Field in Texas. Max Balfour projected that when the school was in full swing, ninety-eight planes would be used and that approximately 45 flight instructors and 200 employees would be hired to run the school. The United States Air Corp of Dayton, Ohio reported that bids were being taken for supplying the air school with 400,000 gallons of fuel and 20,000 gallons of oil.[404]

Cadets assigned to Miami first arrived at Moncton, New Brunswick, just as their Ponca City counterparts had done. Leaving ships and traveling by rail, they found themselves quartered in semi-permanent bungalows equipped with every convenience. They enjoyed the unlimited food and enjoyed shopping in the town, riding the street cars at every opportunity. They were amazed to find shops which stayed open nearly all night and "milk bars" on almost every corner.[405] After their arrival in Oklahoma, and while final preparations were being made in Miami for the British Flying Training students, the new cadets were initially trained for four weeks at the Spartan School of Aeronautics at Tulsa. After only one day to settle in after arrival of the first course, the students were introduced to the Fairchild PT19A Cornell trainer by their new flying instructors. Training began immediately and intensified as time went on. The young men were introduced to solo aerobatics including spinning, formation flying, and finally, night flying. This required a high standard

This aerial view clearly illustrates the training planes lined up along the edges of the runway near the school's hangars.
Courtesy Brian Haslam

The Pierce Pennant Terminal building was reopened after years of lying vacant. After fresh paint and some interior refurbishing, it was used for the administration of Spartan School of Aeronautics in Miami. *Courtesy Ottawa County Historical Society*

ABOVE: This school building in Miami shows the landscaping added after the campus was established. *Courtesy F. Nev Snell*

A large group of the RAF and USAAF staff at Spartan School of Aeronautics. *Courtesy Deryk Maddox*

of skill in instrument flying and it was much more difficult to become comfortable with this new skill. After students moved on from primary training and became senior cadets, they were ready to commence flying with their advanced training in the Harvard AT6. Considered "hot rods," the men enjoyed their flights in this aircraft, although solo night flying was not to be taken lightly in this plane. Paddy Grogan described the flying area around the base:

The land was completely flat save for the man made hills from the debris of the Zinc Mines at Pitcher, a small mining town close by. Roads ran true north and south, east and west, dividing the land up into one mile square sections. This created a grid of roads over which we could navigate almost without the use of a compass. This navigational benefit was splendid until we came to carry out cross country flying away from our area, when we were forced to rely on normal compass navigation.[406]

During the initial period as the British arrived in Miami, the new cadets wore civilian clothing because the United States had not yet entered the war.[407] Most of the RAF trainees were Leading Aircraftmen (LAC) with some re-mustered NCO ground technicians.[408] Course One arrived June 16, 1941 in Tulsa, moved to Miami on July 13, and graduated on October 13, 1941.[409]

Commanding Officers serving at Miami's Spartan School of Aeronautics were Wing Commanders A.C. Kermode, who wrote several excellent operational flying handbooks still being used by teachers today, D.H. Lee, H.A. Roxburgh, A.F.C., and K.N. Sayers, A.F.C., 1945-1946.[410] Mr. Anthony J. Ming, O.B.E., was American Director at Spartan with Chief Ground Instructors F.J. Kiple and W.C. Youngclaus serving him. Chief Flying Instructors were Jack Casey, Al Scarlata, and F.J. Walsh.[411] Remaining personnel are listed in the appendix.

When the schools located in Texas and Oklahoma had need of an expert instructor, they pooled their talents and traveled to the school which needed instruction. In many instances instructors from the school in Ponca City flew to Miami to teach a specific class and the instructors from Miami often flew to Ponca City, probably for the same reason. The flight operations book for No. 3 BFTS documents many instances where staff members from both schools traveled in each direction.

Women played an important role as they became employees at both schools. Sometimes the aircraft maintenance at the school was partly a task delegated to women, who could do all the same tasks as their male counterparts. The shift chief was one of the few men working in maintenance who held a certificate in aircraft mechanics and would oversee the women's excellent work, performed on a regular day shift with a swing shift at night. Seldom did a training mission have to be aborted. Women also served as flight log recorders and parachute riggers.[412] During the period of time that the RAF was in Miami, Mrs. Myrtle Mae Robbins, a homemaker until the war broke out, pressed shirts for the Miami cadets and for the soldiers stationed at Camp Crowder, at Neosho, Missouri, not far from Miami.[413] Lois Shipley was the first woman employed in the flight building at No. 3 BFTS. During her time at the school, she was the associate editor of the *Spartan News* in which the British men in training were mentioned quite often.[414] Another young lady, Patty McNaughton, worked as a clerk-typist in the flight department and became a lecturer on log book procedure to later courses at the school. She was one of the candidates for the most advised member of the dispatcher's staff. She, too, served as a columnist for the *Spartan News*.[415]

Because of the influx of instructors, housing was a problem. For a while, twenty flight instructors and their families were housed in temporary quarters on the seventh floor of Hotel Miami.[416] Miami experienced the same problem as they sought to find good mechanics and other maintenance employees during the time of this school's existence. Paul Grace was the maintenance supervisor for Spartan School, both in Tulsa and Miami.[417]

Although the occasion these cadets were celebrating is unknown, they are pictured wearing aprons and carrying food trays while several women pose wearing the cadet's jackets and caps. *Courtesy Ottawa County Historical Society*

A large group of cadet students listen intently as their flight instructor explains a few of the maneuvers they were expected to learn. *Courtesy Northeastern Oklahoma A&M College*

How the School Would Work

USAAF cadet John Wessale, at the wheel, his wife, Lillian, and their friends, Jack and Dorothy Myers. *Courtesy Kay Wessale Schnieber*

Cadets attending the school were provided clothing, room and board, and $60.00 per month spending money.[418] Spartan School of Aeronautics used the same type of aircraft for training with the exception that the Fairchild PT19 Cornell replaced the Stearman PT17 used in Ponca City.[419] Primary training would consist of ten weeks on the Fairchild PT19 Cornell and another ten weeks would be taught using the Vultee BT13A with advanced training offered on the Harvard AT6.[420] The Skelly Oil Company's president, W.G. Skelly, founded Spartan School of Aeronautics in 1929. Skelly Oil Company supplied the major portion of the aviation fuel used for the school in Miami.[421] The school received supplemental support from Ethyl Gasoline Corporation when Douglas Jeppe, a representative of the corporation, visited the school on July 7, 1942 and presented a lecture on the Internal Combustion Engine.[422]

Other similarities paralleled the Miami and Ponca City schools. Typically, men from both schools were transported in the dark of night through the English countryside as they traveled to Gourock, Scotland. All sorts of armed forces traveled together on the long journey with no food to fill their empty stomachs. Eventually, the men were placed on ships such as the *Queen Mary*. Cyril Grogan, a cadet who would be assigned to Miami on one of the later courses, was in awe of the size of the ship. After a quick look around, Cyril found a magnificent bathroom with hot water, complete with gold plated dolphins for taps. In no time, he slipped into the tub for a quick bath but as he started to soap himself, nothing happened. Alas, poor Cyril needed soap designed for the salt water coming forth from the taps. His exit from the tub was disconcerting as he felt grubbier than when he had entered.[423] The young man later found that he was traveling with the whole British War Cabinet including Winston Churchill. A party of German prisoners of war also was on board the ocean liner. The war cabinet was housed in one half of the boat while the other half contained the troops and prisoners.[424]

Grogan recalled his four-day train ride from Moncton as an uncomfortable journey. The temperature was very cold in New Brunswick but as the train traveled farther and farther south, the temperature became oppressive. Although ice was placed in troughs atop the roof of the carriages, it soon melted and was not quickly replenished. Sleeping also was difficult. Grogan traveled on a train with sleepers. However, no mattresses or bed linens were provided so the young men made the best of it by sleeping on bare boards. When he was awake, Paddy Grogan was amazed at the vastness of the countryside. Much to his pleasure, the entrance of his group of cadets to Miami was memorable, "since the whole of the town's girls, who had just seen off the last course to pass out, were waiting for them to arrive."[425]

Course One Arrives

As Course One arrived in Miami, two months after the signing of the contract made with Max Balfour, the tired young men were welcomed by a large contingency of Ottawa County residents. Forty-nine student pilots arrived by train, although fifty students were scheduled to begin the first course. However, on the day of their arrival, nineteen-year-old Englishman Ronald Dennis Harrison lost his life in a crash of his primary training plane in a field west of Collinsville, Oklahoma. It was August 11, 1941.[446] He was buried in the American Legion section of the Tulsa Memorial Park Cemetery.[427]

USAAF cadet John Wassale and wife Lillian, left, with Dorothy and Jack Wyatt, Miami, Oklahoma, 1944.
Courtesy Kay Wessale Schnieber

The *Queen Mary* transported many of the cadets who eventually trained in Oklahoma. Many pilots kept postcards picturing the vessel. Here, Winston Churchill and a staff member wave to a few of the thousands of young men who shared a ride on the *Queen Mary* as they traveled across the Atlantic on their way to the United States. The ship had a destroyer escort as well as an aircraft carrier escort.
Courtesy Imperial War Museum (A16710)

Primary training at Miami found students using Fairchild PT19 Cornells as they learned to fly.

Courtesy Northeastern Oklahoma A&M College

Mac Bartlett, of the *Miami Daily News-Record*, declared that "the first invasion of cadets was anticipated and that the cadets would take the residents with their engaging smiles, warm handshakes, if with no other weapons."[428] A public reception was held at the municipal swimming pool. After the ceremony, the airmen, unused to eighty- degree temperatures, were allowed to take a dip in the pool to cool off. One Scotsman, used to the hills of his homeland, expressed the observation that he had never been in flat country such as Oklahoma offered.[429]

Transported on a new bus which was eventually dubbed the "Spartan Chariot,"[430] the young men were released from the carriage by a friendly bus driver named "Tiny." The men, who averaged in age from 18 to 21, were besieged with friendly invitations to dances, teas, sports events, rodeos, and church functions as soon as they arrived. Because the construction of their barracks would not be completed for nearly two weeks, the men were temporarily housed in the boys' and girls' dormitories on the Northeastern Oklahoma Junior College campus. The junior college was established in 1924 and continued to carry that name until sometime in 1943, half way through the training of the British cadets in the Miami Community. At that time, the college's name was changed to Northeastern Oklahoma Agricultural and Mechanical College (NEO A&M College).[431] Because school was out for the summer,

The interior of a Fairchild PT19 trainer.

Courtesy Northeastern Oklahoma A&M College

Wing Commander H.A. Roxburgh enjoyed filming new cadets as they enjoyed a dip in the Miami Municipal Pool. The school sponsored competition swimming teams throughout various courses. One of the teams won the first prize of $25 in a Tulsa contest. Funds were usually placed in the RAF Benevolent fund. Several trophies were presented to contestants of Course 25 and 26 by Sgt. Bell, Wing Commander Sayers, and Mrs. McCloskey at one of the Miami swimming galas.

Courtesy Deryk Maddox and Northeastern Oklahoma A&M College

there was no problem in using the structures.[432] Some of the later courses also were housed in the college dormitories from time to time.

Tiny became a standard fixture at the school and remained to provide transportation the entire time the school was in existence. On July 4, 1941, he drove the group for a day's outing to watch the Dewey Rodeo. Later, they were transported on the bus to Bartlesville where they enjoyed cokes, the music of Bob Wills, and a trip to the Frank Phillips ranch.[433]

United States Army Air Force Arrives

The United States Army Air Force cadets of the 322rd Flying Training detachment entered training with the Britons beginning at Course 13 and continued through Course 19.[434] These "Aviation Cadets," as they were titled at the school, consisted of about twenty percent of each of these courses and were enrolled to satisfy some of the Lend-Lease Act regulations between the United States and Great Britain.

As each new course arrived by train in Miami, Oklahoma, the cadets were greeted in the same friendly manner as that experienced by those assigned to Ponca City.[435] Excitement as they began preparing for their future endeavor was obvious. No. 3 BFTS Pilot Officer John Henry Harrad, RAFVR, later to die by enemy action on August 20, 1942, related:

There was from the first a certain spirit which prevailed upon our venture like a following wind at sea, giving our life out there a touch of beauty and dignity. A spirit born chiefly of young comradeship, a common purpose, inter-sympathy and mutual encouragement which was primarily a mark of unique quality and merit to the people and the nation amongst whom we worked and played.[436]

Eric Harrop, a Miami cadet of Course 14, arrived in late October, 1942 on a train called the "Firefly." He was happy to be greeted with warm, sunny weather as opposed to the penetrating cold he had experienced on the earlier portion of his trip from the north. In his diary he described the Spartan campus and mentioned the white timber buildings with green roofs and he went on to write that there was a barber shop, game room, quiet room, reading room, staff quarters, and a gymnasium, all built around a grass square forty-yards across. Beyond the fence at the rear of the complex were the sports grounds containing two football pitches, one soccer field, one rugby field, and a concrete wicket for cricket. One hundred yards from the edge of the soccer field was a railway line and the main flying field, complete with a large hangar.[437] Flying Officer Harrop, of 50 Squadron, was killed with the rest of his crew members of 50 Squadron when his Lancaster bomber was shot down during operations to Karlsruhe, Germany on February 3, 1945. The crew, except for one survivor, was buried at the Choloy War Cemetery in France.[438]

Shortly after their arrival, No. 3 BFTS adopted an official song. The words to *Ad Astra* were written by the staff and cadets at Spartan, while the music was written by R. H. Cort. The melody came from a tune originally composed by Mr. Cort for the RAF Apprentice Training School.[439]

This pen and ink drawing of downtown Miami was created by Cadet J.H. Widgery and featured in a 1944 *Open Post Magazine*. Cadets were excited as they viewed the bright lights of Miami after two years of black-out in England. *Courtesy Northeastern Oklahoma A&M College*

Several cadets explored a variety of shops in downtown Miami.
Courtesy Northeastern Oklahoma A&M College

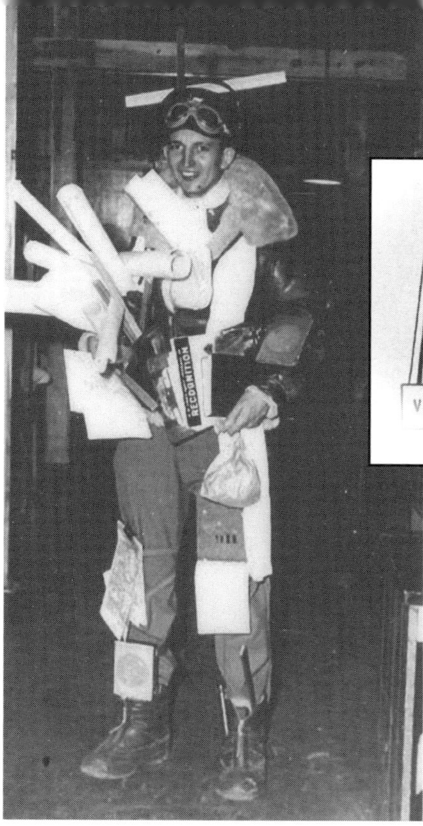

Per Ardua Ad Astra was the official logo used by No. 3 BFTS in Miami, Oklahoma. *Courtesy Ponca City Library*

ABOVE: John Wessale, Course 19 USAAF cadet, struggles with maps as he prepares to pass his wings exams. Wessale was ranked 6th in his class when he was awarded his flying badge. The order of merit was based upon marks awarded for flying, ground examinations, and for positive qualities exhibited as an officer or non-commissioned officer. *Courtesy Kay Wessale Schnieber*

Ad Astra

How mighty the Spartan who strove to attain to
Perfection of manhood in contests of old!
May we who endeavor his precepts to follow
Be steadfast in danger, in battle as bold –

Chorus:
So open the throttle and join in the chorus
The blue sky above and the earth down below;
The thrill of the air, of adventure before us,
Our hearts ever eager to combat the foe.

If like the Spartan, we proudly determine
To reach our objective, and all that it brings,
How proudly we'll wear on the breast of our tunics
The prize that we've worked for, the coveted "Wings."

Chorus:
And when we maneuver our Spitfires and Blenheims,
And chase the intruder from out of the sky,
Old mem'ries will call us to scenes of Miami,
In far Oklahoma – where we learned to fly.

Cadets queue up at the school's PX, where they could purchase a variety of candy and other necessary items such as writing materials and grooming aids. *Courtesy Northeastern Oklahoma A&M College*

Groups of cadets took their march to the cafeteria seriously as they prepared for each meal. *Courtesy Ponca City Library*

Day to Day Routines Take Shape

As each new class arrived, the course was divided into flights with a cadet in charge.[440] A typical day for the students began as they got up at 5:45 each morning. After a series of calisthenics, the men marched to breakfast in crisp, military cadence with arms swinging stiffly. After breakfast was served, the young cadets returned to their barracks where they straightened their rooms. In the afternoon, while half the students received ground school training in the classroom, the other half received flying instruction. The schedule was reversed the second half of each day. The men marched to the cafeteria for lunch and held a "dinner parade" each evening. Many evenings found the American Army Air Force flight trainees being trained in close-order drill after their marching skill was found to be a little too casual. Games were enjoyed three times a week and a study hall was provided in the evening. Eight hours of sleep was required each night.[441]

Few problems surfaced as the young men went about their daily routine but Course Four was one which could not brag about its luck. Things seemed to be going well until it began to rain, and rain, and rain. The men quickly adapted and found plenty of things to do indoors. They had sing-song and smoking concerts and played chess, darts, and table tennis while they waited for the go ahead to continue flying. They made up for lost time by flying on Saturdays and Sundays but lamented the large amount of rain in their school magazine:[442]

Rain

When we're back in dear old England
And we hear the gentle rain,
With its pitter-pitter-patter,
Beating on the window pane,
Let us silently remember
That we ought not to complain,
For it never rains in England,
Quite like Oklahoma rain.

Oklahoma weather was very difficult for the students to become acclimated to. The heat was especially unbearable to them and one student wrote his parents telling them that even at four o'clock in the afternoon, the temperature was 93 degrees. He tried to solve the problem as he sat in front of an electric fan with just a pair of shorts. However, even the fan blew hot air on him.[443]

Food was on the top of the list of things which these young men enjoyed once they arrived in the United States. The "Will Rogers, Ltd." offered a menu advertising a special RAF breakfast for 75 cents. It included a choice of fruit juice, cereal with half and half, two eggs with broiled ham or bacon, buttered toast and jelly, along with coffee, tea, or milk.[444]

Retired RAF Air Commander J.F. Langer reiterated these feelings for the wonderful food he experienced when he arrived in Miami. He loved medium-rare T-bone steaks with an egg cooked sunny side up, served with a heaping portion of French fries, and topped off with apple pie and ice cream. He and his friend "Spud" Murphy also were treated to barbeque by Ma and Pa Black of Baxter Springs, Kansas.[445] All the cadets enjoyed their jaunts to town to visit the local soda fountains.

Once in a while members of No. 6 BFTS in Ponca City would travel to Miami in order to pit their skills against the men in Miami as they played in exhibition soccer games. Most often Ponca City lost after riding in an uncomfortable school bus on the trip to Miami. The men of No. 3 BFTS then endured the same five hour drive in an uncomfortable yellow school bus when they traveled to Ponca City for a game. The Miami cadets described it as "a novel and successful form of offensive sweep upon British cadets at Ponca City."[446]

Water sodden fields made it impossible for cadets to get into the air as they waited for better weather and the airfields to dry.
Courtesy Kay Wessale Schnieber

A group of British cadets gaze at themselves in the mirror behind a local Miami soda fountain as they enjoy a variety of delights.
Courtesy Northeastern Oklahoma A&M College

Some Miami cadets also put on soccer exhibitions at the local stadium and in the surrounding towns. They played against the 38th Battalion Signals Regiment at Camp Crowder near Joplin, Missouri and won the game 8-0.[447] Basketball became a favorite sport for the men in Miami and they held several inter-flight competitions. After several games between themselves, they played at the Miami High School and although they lost, more than $100 was raised for the Infantile Paralysis fund.[448] Both schools had cadets who enjoyed boxing. In Miami, the first amateur boxing team from No. 3 BFTS went to the Northeastern Oklahoma A&M Junior College to box. They won four and lost five.[449] A few men enjoyed inter-flight games of American baseball, football, tennis, and basketball while were they were in Miami and competed with Miami high school teams.[450]

Bert Sims, top far left, enjoyed playing soccer with his companions while he trained in Ponca City. Many exhibition matches were offered to the public while both schools were in existence. *Courtesy Albert Sims*

Active cadets took their physical training seriously as they prepared for the war in England. Closer to home in Oklahoma, No. 3 BFTS sometimes took on No. 6 BFTS in exhibition soccer matches and quite often they were the victors. *Courtesy Ponca City Library*

Miami Enjoys the Service Center

A suitable gathering place for the young men assigned to Spartan School of Aeronautics became a problem not long after they arrived in Miami. At first, citizens began to invite small groups of men into their homes but as the war progressed and rationing of gasoline and food became a problem, it became increasingly difficult to entertain the large group of cadets who wandered the town during open post. The "Miami Forum" was organized by a group of concerned citizens after the Dobson family donated the use of one of their downtown buildings rent free for the duration of the war. Within a week, the Forum had organized various civic groups and churches and had quickly furnished the service center. It officially opened on Thanksgiving Day, 1942 and operated on a yearly budget of $1,500. The young men enjoyed their time by reading, playing games, and dancing with hostesses who were especially selected by clergy and reliable citizens of the community. These young women donated their time one evening a week to entertain the servicemen. The center was open from ten in the morning until ten in the evening and strict rules were enforced. The cadets were not allowed to make dates while they were in the center and the young women were not allowed to be escorted home by the young men. Many young men expressed their gratitude for the donation of the center and felt it was a warm and cozy place to spend their free time.[451] While they were there, the cadets also listened to the American radio networks with the "big band" sound of Glen Miller while they enjoyed the service center or the USO.[452]

When the Miami cadets wanted to roam during their free time, they were often invited to Camp Crowder in Neosho, Missouri. Many evenings were spent at dances in the camp's gym after invitations were issued from the camp's officers. Camp Crowder had a band which performed under the direction of Lieutenant Franco and on October 31, 1942 it was reported that the dance given by Camp Crowder was "most successful and the best dance music ever had at the school."[453] The latest Hollywood movies also were enjoyed on the base. The English boys took great delight in eagerly showing their prowess at soccer as they played against a less experienced team from Camp Crowder. Although a deluge of rain produced a quagmire on the playing field, it added to the amusement of the spectators, who also laughed as Fort Crowder players asked for baseball bats to finish the game. They also got a good laugh as they lost their baseball caps, only to lose 0-8 to the enthusiastic Englishmen.[454]

Hostesses at the Miami Service Center were often seen with a friendly smile as they passed around plates of home-made goodies. The treats were appreciated by the Miami cadets as they spent time at the service center.
Courtesy Ottawa County Historical Society

On other occasions, the Joplin, Missouri United Service Organization and musical artists, including the Fritz Newman string quartet, performed for the cadets in special concerts. As the USAAF cadets graduated, they went to Camp Crowder to purchase their officer uniforms. They had been given a sum of approximately $250.00 to pay for both winter and summer uniforms, and accessories such as shoes, socks, and underwear.[455]

Camp Crowder, a training center for the United States Army Signal Corp, was built in 1941 and ultimately had nearly 47,000 troops stationed there during the war. The popular cartoon, "Beetle Bailey" was created by Mort Walker, who used his experience while based there when he created "Camp Swampy." The camp closed in 1946 as a training base.[456]

In Miami, the Coleman Theatre always was busy as the cadets arrived after walking into town from the school. On Wednesday evenings, the movies were free and tickets were given to all who attended. Sometime during the evening there was a drawing with the holder of the lucky ticket stub receiving fifty dollars.[457] The men often brought a date. They consumed volumes of popcorn and sodas. The Brits also attended movies at the Glory B but, according to Eric Harrop, it was "nothing but a flea pit."[458] *Weekend in Havana* was one of the movies featured in Miami as the first group of men arrived in that city.[459]

In Course 19 the cadets submitted photographs of sweethearts and wives which were judged and given the title "Miss 19 Course." In that particular contest, more than thirty cadets submitted their sweet hearts' photographs.[460]

One of the first movies enjoyed by the new cadets just after the opening of the school in Miami was *Weekend in Havana* at the beautiful Coleman Theatre. *Courtesy Ponca City Library*

ABOVE: The Coleman Theatre, established in 1929, was originally named the Coleman Theatre Beautiful. The Spanish Mission Revival styled theatre was built by George L. Coleman, Sr. Former No. 3 BFTS cadets still return to visit the structure. *Courtesy Coleman Theatre*

This Miami street scene pictures the Glory B movie theater. Joel McCrea and Ellen Drew starred in *Reaching for the Sun* in this 1941 film. *Courtesy Northeastern Oklahoma A&M College*

On many Saturday evenings Miami cadets could be found at one of the local drug stores where they enjoyed a variety of luscious ice cream concoctions. Later, after a movie with their dates, the men would hitchhike back to the air field where they continued to enjoy telephone calls from many girls and eager town folks who would issue invitations to their homes for the following day. They had phenomenal success thumbing rides to and from school and they were often given rides by bootleggers. It didn't take a wizard to realize what the men were selling because the sound of bottles clinking in the back of the car was easy to hear. Prohibition was the law and Oklahoma was a "dry" state, whereas Arkansas and Missouri were not. Therefore, they were very popular destinations when the men went on leave.

Eric Harrop would arrive at the Hotel Miami on Sunday mornings, where he was always invited home to dinner by one of the local families.[461] Later, if he did not have a ride, he would hitchhike back to the school.

Another Miami cadet penned these words:

Did You?

Did you give him a lift? He's a brother of man,
And bearing about all the burden he can.
Did you stop when he asked you to give him a lift?
Or were you so busy you left him to shift?

Did you give him a smile? He was downcast and blue,
And a smile might have helped him to carry it through.
Did you ask what it was, why the quivering lip
And the glistening tears down the pale cheeks that slip.

Did you bid him good cheer? Just a word or a smile,
Was what he most needed that last weary mile.
Did you give him a hand? He was slipping downhill
And the world so he fancied was using him ill.

Do you know what it means to be losing a fight
When a hand just in time would set everything right?
Do you know what it means just the clasp of a hand
When a man's borne about all a man ought to stand?

Oh! I know what you think – and it may be quite true,
But the test of your manhood is – what did you do?
Did you help him along – Did you show him the road –
Or did you just let him go on with his load? [462]

Two unidentified Miami cadets marvel at the large fields of cotton growing in Oklahoma. *Courtesy Northeastern Oklahoma A&M College*

Spartan School of Aeronautics devised a plan which helped all cadets to enjoy invitations to dances and other parties from citizens of the surrounding areas. Thirty cadets were invited to fifteen homes one weekend in 1942 and enjoyed a dance at the Chanute, Kansas Country Club. Their pretty, wholesome, and friendly young dates wore beautiful formal gowns to the dance. The next morning the youth enjoyed swimming and golfing and later assembled in mass at the Town Hall where they were guests of the American Legion.[463] The month prior, another thirty cadets were treated to dinner and a dance by families at Neosho, Missouri.[464] Other dances were given in the towns of Baxter Springs, Fredonia, Galina, Columbus, and Pittsburg, Kansas for the benefit of the gentlemen. Most dances were heavily chaperoned[465] but when the Miami cadets found their way across the state line in Joplin, Missouri, they plied their dancing skills at The Rendezvous, a local night club.[466] At other times they danced at the USO in Joplin, Missouri.

Miami cadets were issued an invitation to attend a Quapaw Stomp Dance at the home of Alexander Beaver. They marveled at the rhythm of the drums. Many photographs were taken to mark the momentous occasion and to record a gambling game played with colored beans. The following week the chief, Victor Griffin, along with his family, were entertained at the flying school where they were served lunch and were given a tour of the facility. They were most impressed by the gyrations of the Link Trainer.[467]

One of the most enjoyable events that some of the young cadets of Course Ten experienced while training at Miami was their participation in a donkey baseball game held in the Miami High School parking lot in September, 1942. After several exhibition innings, the cadets were invited to play a game against the Miami Junior Chamber of Commerce. Each member of the team, except for the pitcher and catcher, was mounted on a donkey. As they struck the ball, the runner had to ride his donkey to each base. One of the English players was soon to discover that it was not as easy as it might seem as he tried to persuade his sturdy mount to run to base, only to be seen "cantering at the rear of the stadium." The RAF cadets won the game eight to one.[468]

Many families, including children, invited the Miami cadets to Sunday dinner and enjoyed other activities with them. Lifelong friendships often resulted. *Courtesy Northeastern Oklahoma A&M College*

An unidentified British cadet stands in front of a downtown building. It is a typical scene enjoyed by the citizens of Miami and Ponca City. Often these men hitchhiked to and from town on leave.
Courtesy Mariann Bryant

ABOVE: Officials from No. 3 BFTS enjoyed their visiting dignitary, Chief Victor Griffin, of the Quapaw Tribe. *Courtesy Ponca City Library*

LEFT: Chief Victor Griffin inspected a Link Trainer when he and his family visited Spartan School of Aeronautics. *Courtesy Ponca City Library*

Ron Emmerson arrived on the North American continent via the S.S. *Mauretania* and after his initial time spent in Moncton, New Brunswick, left the No. 31 personnel dispatch center and traveled to Spartan School of Aeronautics in Miami. Growing up as a child, he had served as a young male soloist and eventually became a tenor in his home town. He had also learned to play a variety of musical instruments including the piano, pipe organ, clarinet, saxophone, and bagpipes during his young life. Soon after his 1944 arrival in Miami, he sought out the local Presbyterian Church and immediately became part of the choir where he sang with a sextet. Not only did he and other cadets attended Sunday services but participated in other church activities as well. Ron had married his wife, Emma, just before he left for Miami. While Ron was in Miami his first child, David, was born. Immediately after his birth, the Miami church sent Emmerson congratulations upon the birth of the boy. Many lovely baby clothes which were not available in England also were sent by many of the parishioners as gifts to the family. The pastor at the time was Reverend Harry Curtis.[469]

The First Presbyterian Church of Miami, Oklahoma printed the following in its Sunday, March 19, 1944 bulletin:

The Air-Man's 23rd Psalm

By Robert Cashman

The Lord is my pilot; I shall not want. He maketh me to rise up into the heavens. He leadeth me through clouds of star dust. He restoreth my soul. He leadeth me in paths of wonderment for his name's sake. Yea, though I fly through the dangerous and sickening air pockets, I will fear no evil, for Thou art with me. Thy stars and Thy moon they comfort me. Thou openest lovely vistas before me in the presence of the angels. Thou fillest my heart with delight. My cup runneth over. Surely goodness and mercy shall follow me all the days of my life, and I shall fly in the clouds forever. [470]

Time Off–A Welcome Respite

As cadets completed their flying requirements, leave was usually granted for short periods of time. Many of the men often stayed close to the school while enjoying their time off. Mr. and Mrs. J.R. Blaine of Disney, Oklahoma, were known to take many of the Miami cadets on short boat trips on Grand Lake in their boat, the *Evening Star*. [472] Others ventured a bit farther from the school. Eight students of the No. 3 British Flying Training School in Miami were guests of Mid-Continent Airlines, Inc. on one of their sojourns to Kansas City. The airline took the men on a 15-minute flight over the city while they viewed various points of interest. After they landed, the Kansas City Rotary Club continued the sight seeing adventure. [473] Back in Miami, the well known Dobson family provided a Service Club for the young men and it was often visited when they could not travel or when they only had a few hours before they were required to return to the school. [474] When they were able, the cadets from No. 3 BFTS also enjoyed ventures to Joplin, Missouri, Baxter Springs, Kansas, Tulsa, Oklahoma, and Kansas City, Missouri. [475]

The First Presbyterian Church of Miami, Oklahoma welcomed Ron Emmerson, front row left, to their church choir while he trained as one of the cadets at No. 3 BFTS. At the All Saints Episcopal Church, 16 other cadets performed for the Episcopal Bishop of Oklahoma as he visited the church. Voluntary church services and church parades were also held at the Spartan Campus and at local churches. [471] *Courtesy Ron Emmerson*

Cadets enjoyed the Neosho River and the Grand Lake of the Cherokees during their time in northeast Oklahoma. *Courtesy Northeastern Oklahoma A&M College*

Pleased with the hospitality and entertainment often provided by local families, one class of students decided to return the favor and on Tuesday, December 15, 1942, a crowd of 500 Miami guests enjoyed a performance by the cadets. A drama of murder and justice called *Rope*, written by Patrick Hamilton and performed in the Miami High School auditorium, was met with a great deal of pleasure.[476] One hundred thirteen dollars was collected and divided equally between the Royal Air Force Benevolent Fund and the Miami Service Center, the equivalent of a local USO.[477]

"Mama Jenks" visited the northeastern Oklahoma town[478] in April, 1944 when she entertained many young men from Course 20 at No. 3 BFTS. Most often she and members of her family would hold a special event such as serving a meal at a local hotel. Some of the cadets thought so much of her that they presented her with a RAF jeweled wing pin on Mothers' Day when they thanked her for her hospitality during their stay in America and for all the good times they had shared in Kansas City.[479]

Several RAF cadets say farewell to some of their friends as they board a train during one of their leaves. Men from both Oklahoma schools sometimes rode the train to Colorado to visit American host families. Often their passage was discounted by the railroad company during their time in the United States. *Courtesy Mariann Bryant*

BELOW: Donkey baseball seemed like a good idea when these two cadets joined the game held at the Miami High School. They later won the game 8 to 1 against the Miami Chamber of Commerce.
Courtesy Ponca City Library

ABOVE: Many cadets took the opportunity to travel during their leave. These cadets stopped at the Oklahoma/Texas border and recorded their visit to the neighboring state. *Courtesy Mariann Bryant*

Lucille "Mama" Jenks held a special place in the hearts of her adopted British sons, many of whom visited her family home in Kansas City. Two cadets in this group, Tom Downs and Pete Dowthwaite, were in the last course to train in Ponca City. When Darr School closed in April, 1944, these men finished their training in Miami. Others were transferred to the remaining American BFTS schools. *Courtesy F. Nev. Snell*

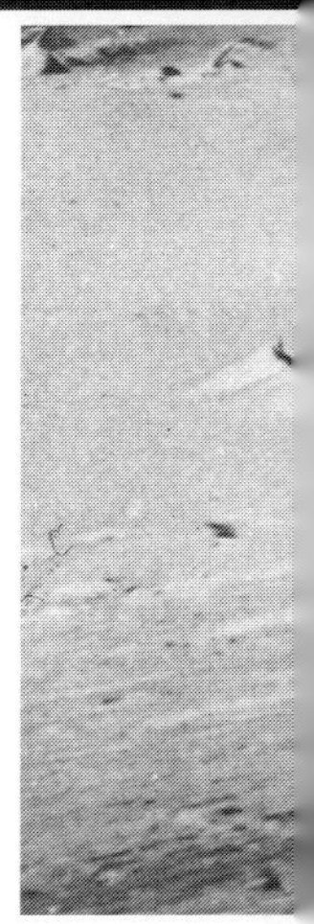

It would have been difficult to prove that the cadets who remained in Miami during the Christmas season of 1944 were unhappy to be so far from home. Course 22 decided to hold a Christmas party at Northeastern Oklahoma A&M College. The evening started with Christmas carols performed by the school choir and later in the evening the party moved to the Spartan cafeteria for dancing. The band was composed of cadets from the school. During their performance there was an invasion by "two barbarians from Scotland, complete with kilts and 'a wee drap o'whusky'." The "Five Drips" then arrived to perform "The Shower," accompanied by a guitarist and groaners, also from No. 3 BFTS, who kept the entire audience in an uproar of laughter. Course 22 said they would remember their holiday for years to come.[480]

As each course progressed through the detailed ground school courses they were required to take numerous written examinations. They spent more and more time in the air, completing the required number of hours of flying time. In addition, night-flying operations were required as part of the primary flight segment, as were solo night landings. These were dangerous, but most students managed to get through

The Grand Canyon piqued the interest of two Ponca City cadets during their time away from Oklahoma. Although it was hot and difficult to reach the bottom of the canyon they enjoyed the hike down the steep, rocky trail. *Courtesy Dr. and Mrs. J.E. Fuller*

Cadets enjoyed the novelty of an Oklahoma snowfall during their training in Miami. The snow was headline news for the local hometown newspaper. *Courtesy Ponca City Library*

their maneuvers and their courses without too many difficulties. Instructors were eager for the young men to do well and if one felt he was not helping the student enough, students were assigned to another instructor whose teaching style might be able to make the difference.

Later, in advanced training, cross-country flights of 300 or more miles were required. Two students would fly on these cross-country maneuvers, taking turns as each flew a leg of the journey, which was usually flown in a triangular course. On August 31, 1942, a flying accident occurred one mile northeast of Chanute, Kansas. Two Vultee BT13A aircraft collided in the air while on cross-country flights. LAC cadets Alan Brown, Harold Alfred Burman, Herbert Horace Hacksley and Donald Arthur Harfield were killed in the incident and were buried several days later at the G.A.R. Cemetery in Miami, Oklahoma, under the direction of the Cooper Funeral Home. Several officials attended the ceremony.[481]

This cartoon drawn by a Miami student from Course 13 jokingly portrays the fear the men had of performing foolishly while flying and their dread of being eliminated, or "washed out." *Courtesy Ponca City Library*

All students were required to finish all their hours in the air and each of the men kept a log book, meticulous in every detail. Immediately after graduation examinations, the men who passed were usually given several days leave before returning to school to begin another section of their studies. The students traveled to many locations, with several even reaching Hollywood. Some of the cadets used their servicemen's passes for discounted tickets on railroads and buses while others hitch-hiked their way across country. As they hitch-hiked, they were tremendously impressed by the vastness of America and the curious and friendly nature of her citizens.

Not everyone made it all the way through a course. Although it was a sad time for the young cadets who did not pass their examinations, became ill, or for other reasons were not able to complete training for their flying skills, it was necessary that they be reassigned to other locations for training to benefit the war effort. A few young men, even though they did well in their studies and flying, disobeyed the school's rules for inappropriate flying and as soon as they were observed, they, too, found their way back to Trenton, Ontario, Camp Kilmer, New Jersey, or Moncton, New Brunswick.[482]

One of the students who graduated from No. 3BFTS later told of his near miss in "washing out" from the school. He and his friend were flying together over Grand Lake near Miami on one of their training flights. Ducks were swimming near the shoreline and the two cadets thought it would be great fun to "harass" them by swooping low and causing a great roar. Immediately the ducks began to scatter as the rude young fliers beamed in on them several times. On their last pass, the men came so near the lake that their propeller barely touched the water. It was enough to damage the airplane and although they could fly the aircraft, the cadets knew they were in serious trouble as soon as they landed back at the field. Desperately they attempted to come up with a reasonable explanation but since they could not think of one, they decided there was only one thing left to do in order to save themselves an early trip back to Canada.

The radio control tower was informed that they were preparing to land and after several seconds, the men falsely reported that they could not get their landing gear to drop. The control tower operator instructed them to try again for several times while the men circled the field as they pretended to do what was requested. Finally, the operator told them that they would have to land on the belly of the plane and to come on in. The two cadets managed a reasonably safe landing but

The *Open Post Magazine* often held humorous stories and jokes within its pages, exemplifying the sentiments of many of the young men toward the local girls and likewise, the girls' thoughts:

Jane, seated at one end of the sofa speaking to the cadet seated at the other end:
"George, do you think my eyes are beautiful?"

Bashful George: *"Ra-ather!"*

Jane: *"And do you think my hair is the prettiest you've ever seen?"*

George: *"By Jove, it is!"*

Jane: *"Do you think I have a perfect figure?"*

George: *"But definitely!"*

Jane: *"And do you think my lips are like rubies?"*

George: *"Mmm."*

Jane: *"Are my teeth like pearls?"*

George: *"Absolutely!"*

Jane: *"Oh, you English say the cutest things!"*[483]

with the body of the airplane dented in several places. No one was the wiser and, of course, they graduated while the aircraft was repaired at the expense of the unsuspecting school.

No matter where the British students wandered, whether in the towns where their schools were located, or wherever they seemed to migrate on leave, differences in American language and that used by the Brits always was noted. Terminology of common items produced laughter on both sides. For instance, what Americans called cookies the British called biscuits. As the British boys would say "good bye," they were very much stymied when the response was "Ya'll come back again." The game of American checkers was called draughts by the English and trucks were called lorries. Standing in line was queuing up, gasoline was called petrol and candy was called sweets.[484] To add to the humor, an American tea bag was called a dead mouse by the Britons and a yard or lawn in America was called a garden in England.[485] In England a yard would have been somewhere where many buses were parked. The list went on and on.

English cadets gathered around their Link Trainer instructor to learn the many procedures necessary for them to learn to fly.
Courtesy Ponca City Library

This beautiful young woman posed with three of her British friends at her home in Miami. She is typical of the friendly local beauties which captivated the hearts of the young cadets during their stay in Oklahoma. *Courtesy Ottawa County Historical Society*

About a year after Spartan School of Aeronautics started training British cadets, the patrons of Tulsa took up the welcome task of entertaining thirty of the cadets. The men were entertained for a weekend in Tulsa as guests of members of the American Association of University Women. It was their feeling that:

Real friendship and interest in the boys themselves will do much to promote good feelings between our nations. [486]

Members of the Panhellenic at the University of Tulsa began the festivities by serving a buffet supper. They were then taken to a Tulsa football game. Following the game, they were entertained at another party and finally retired as guests in the homes of a variety of citizens. After sleeping late the next day, the fellows were off to visit radio station KVOO. Later, they enjoyed ice skating and a trip to the Philbrook Museum of Art. [487]

Not everyone experienced the same things that weekend in Tulsa. Some of the cadets, who had never seen an oil well, were taken out to an experimental well which was being drilled nearby. They enjoyed investigating the derrick and other machinery. In one of the homes of their host and hostess, several young men saw their first electric dish washer. The attic fan also provided amusement for the men who had never seen one before. [488]

During the weekend, some of the hosts made plans to take the cadets to see the mountains of Colorado and a trip to Golden and Denver was planned for a later date. [489] The word spread throughout both schools that there were congenial people in Colorado Springs who would be happy to host them during their travels. Soon, several groups from both schools were entertained by Mrs. Hazel Abbot of Colorado Springs, who also hosted several young men from No. 6 BFTS at Ponca City. [490]

Pete, one of the camp mascots at Miami, was decked out in his own specially designed uniform provided by No. 3 BFTS cadets. *Courtesy Northeastern Oklahoma A&M College*

Another Tulsa group took every opportunity to help the cadets from Miami and Ponca City. "Bundles for Britain" was organized at Christmas time in 1942 when it became a placement bureau and headquarters for the RAF cadets. They were located at 1309 East 15th Street and the organization saw to it that the cadets were placed in the homes of Tulsa hostesses where they were provided with room, board, and entertainment during their stay in Tulsa. At times, the organization also hosted parties and dances for the men. As the word grew among the British boys that they were being treated well, larger groups came into Tulsa each weekend. Self-addressed cards were sent to both schools requesting that the young men fill out an information form when they knew they were coming to Tulsa. This allowed the chairman to place them in the various homes for the weekend, and sometimes longer. Financial support for the organization was gained through individual contributions, although a local alumnae group of Pi Beta Phi sorority became the first group to make a collective donation.[491]

Missing their canine friends at home, the young men adopted several mascots at the schools. Not long after their arrival in Miami, some new cadets noticed a stray dog hanging around the barracks which appeared to be a German Shepherd. Inquiries were made and it was soon discovered that the dog, originally named "Scout," was owned by Mrs. Frank Charleton. She quickly relinquished claim to her beloved dog when it became quite apparent to her that LAC Fisher of Course 6 loved the dog and it loved him in return. The dog was renamed "Pete." His daily routine started as he paced in front of the barracks, waiting for the first group of men to walk to breakfast. As they entered the cafeteria, Pete quickly ran for the back door where he always received a hand out. Later in the morning, after a quick nap, he spent his time chasing cars and nosed around the school while he waited for the flight line to return for the evening meal. After his meal, he liked to accompany the guards as they made their rounds of the school. Pete became a fixture around the school.[492]

At Wings Parades, Pete was outfitted in a red and black lettered blanket and sometimes appeared carrying a clan tartan donated by some of the young men from Scotland or Ireland. After his death, a kind-hearted citizen of Miami quickly provided a new pup, a German Shepherd police dog. The men quickly named him Re-Pete.[493] Flake was another of the beloved dogs taken in by the cadets before the school closed in 1945.[494]

Miami won the young men's hearts. One Miami cadet said:

We met America not in its politics or business, but in its restaurants, cinemas, hotels, shops, drug stores, saloons, dance halls, churches, ball games, rodeos, country clubs, hot-dog stands and railroad stations. We were absorbed into American life, if at first curiously, then easily after the first differences of the two customs had smoothed in contact.[495]

Many adopted Oklahoma "moms" found they had plenty of help in the kitchen as they used their own families' rations to prepare home-cooked meals for the hungry young cadets. *Courtesy Dr. and Mrs. J.E. Fuller*

Deryk Maddox, far left, with his friends in front of their aircraft in Miami. These men were representative of all those whom Mrs. Hively described as they visited her home in Commerce. *Courtesy Deryk Maddox*

Two Miami cadets enjoyed swinging their pretty companion as they enjoyed their visit to one of the many hosts who entertained them during their stay in Miami. *Courtesy Ottawa County Historical Society*

Mrs. Hively Loved Her Boys

Mrs. Margaret Hively, of Commerce, Oklahoma, reminisced about her British guests when she wrote her memoirs of No. 3 BFTS in 1943:

It is a good four thousand miles from the "tight little island" of England to the wide open spaces of our great southwest. And that is some hop in any man's experience. But hundreds of young Englishmen have made the trip in the past three years. They are the cadets of the Royal Air Force being trained in various flying schools in this country.

These boys are chosen from all ranks of life in their native land – from aristocratic manor houses and from London's east side. But most of them come from comfortable middle class families such as comprise the bulk of the United States population.

On Sunday afternoons I pick up some of the boys at the air school and take them home for Sunday night supper, or "tea" as they call it. If it is winter, they will be wearing the dark grayish blue uniform of the RAF but during our hot summers in Oklahoma, they are issued the regulation khaki that American eaglets wear, retaining only the wool overseas cap, and the inverted wings of the RAF on their khaki sleeves. They have the proverbial English complexion – smooth and rosy, and the quiet courteous self-controlled English manner. But under the sunshine of American kindness, they visibly relax and chatter with the frank spontaneous gayety of youth the world over and show a zestful, enthusiastic curiosity for our mode of life, thoughts, habits, customs, flora and fauna, - everything labeled "American."

Their first and deepest impression of America is always the same – an impression of bigness – both in the area and in the hearts of the people here. They are agreeably surprised by the attention they attract. People stop them with kindly questioning on the streets of the cities and even occasionally inviting them to speak over the radio. "Some tosh!" a pink cheeked youngster confided to me. "We feel very much flattered. In England, soldiers are so common and of so many nationalities and countries, our good points are overlooked." He grinned engagingly. Who says the English have no sense of humor? That's pure baloney!

"One thing I particularly notice here," said another twenty year old, "is the clean spic-and-span newness of everything. The fresh painted wooden cottages with red and green and blue roofs. The broad paved streets, and lawns stretching the full length of blocks, without fences. It is because we are so crowded, I suppose." The fence answers an unconscious desire for privacy.

They like the arrangement of my house, too. The big living room and dining room are practically one and they take a delight in the various household gadgets such as electric mixers. "In England," they say, "Only hotels, restaurants and the very rich have electric ice boxes."

All this is interesting and somewhat amusing to me but when I mention that I am an antique hound and trot out my precious collection of old pressed glass, my grandfather's Confederate cavalry saber, the old muzzle-loading shotgun with which he killed deer right her on Cabin Creek, and as a piece de resistance, the old Pennsylvania Dutch clock with the wooden wheels, it was their turn to be amused for in England, it seems they don't begin to reckon anything as "old" until it is at least three or four hundred years. The church that one cadet attended in Stratford-on-Avon was built in the eighth century.

When I first knew these young flyers I took pains to have their national drink brewed fresh and strong whenever they came to "tea." I soon discovered that not all Englishmen prefer tea to coffee. "You see the reason we drink tea instead of coffee in England," they explained, "is because we make good tea but we don't make good coffee as you do here."

They like iced coffee, too, but I've never succeeded in putting over iced tea, or hot tea with lemon. The real preference of most of the lads is milk, milk, and more milk!

Along with the milk they like fresh eggs. They help me feed my flock of hens and vie for the privilege of gathering the eggs. "If Mother could see me now!" laughed on boy, holding up four eggs in one hand and three in the other. "Seven months' rations in my hands, all at once!"

As I buzz about the kitchen preparing supper, the boys follow at my heels, telling me of their homes in England and what they do in peace times and what they plan to do when the war is over (Many want to come back here!) Wallets are brought out and treasured snap-shots are exhibited with boyish pride. "This is my home. That big bush you see is a lilac. It is very, very old. Its trunk is a big as a tree trunk and in spring you can smell the blossoms for blocks." And – "This is my dad and little sister. I'm paying for piano lessons for her." And – "This is Marguerite, my girl. She works in the City." (Equivalent to our civil service.)

In spite of their manly reserve, I gather that the boys get a little homesick and their thoughts dwell a great deal on the folks at home. This is especially noticeable when they partake of some American bounty such as fresh fruit, ice cream, or chocolate candy. At such times they invariably express the wish that the loved ones back home could share in these good things. Speaking of candy, you should see the embryo English falcons when I suggest we make fudge. Does that get over big! Such quantities go down the smooth young throats that I am in a momentary panic concerning my sugar coupons and begin to cast around in my mind where I can beg, borrow or steal more of the makings for this top treat. "In England, candy is out for the duration," they say. "Even in peace time we don't have the variety of bars you have. Ours are mostly slabs of sweet chocolate with sometimes raisins added."

The lads took to fried chicken and hot biscuits like ducks to water. Scones, they call the biscuits. And their biscuits are what we call cookies but scones, or biscuits, or whatever, the hot buttered bread disappears at the average rate of a round dozen per man nor do they show slouch work disposing of buttered toast (margarined, of course) thickly spread with plum jell.

After tea is over, I wash the dishes and the boys very carefully dry them. Then we go outside and wander about the orchard, under the apple and peach trees. At my invitation, they help themselves to any ripe fruit. "Um-yum!" Strong white teeth bite into the soft scarlet of the Red Bird Clings and pink juice drips from eager young fingers. "Imagine picking ripe peaches from the tree! We have peach trees and grapevines, too, but the fruit never matures. Now, apples – that tree over there looks exactly like one in our back yard." When not one more bite can be tucked away, a paper sack is filled for their delectation through the coming week and we retire to the lawn swing. There we amiably discuss all manner of topics – the differences and similarities in our Congress and their Parliament – our respective school systems – the war – the robot bombs – our identical dislike of German and Japanese tyranny and cruelty – and countless other things.

At nine-thirty the lads say goodnight and cheerio. They must be in their barracks by ten. As I climb the stairs to my bed, I marvel at the great basic likeness between the English and American people. Our English cousins are truly kinfolk, indeed! [496]

Miami Losses

There were fatal accidents at Spartan. Fifteen British chaps are buried side by side in the Grand Army of the Republic Cemetery, less than one mile north of the flying school, which is located on US Route 66 in Miami. [497]

Wally Elliott, a RAF pilot, was killed while flying on a cross-country solo night-flight. His aircraft's wreckage was found in the middle of State Highway 96 nine miles west of Columbus, Kansas, with Wally still in the cockpit at the time of the crash. His aircraft had dived into the ground. Considered old by the younger pilots who averaged 19 to 22 years, Wally was 29 when he departed life. A military funeral and parade was held with all members of his class marching in a slow cadence, pausing after each step. The Reverend Harry Curtis of the First Presbyterian Church conducted a service on October 24, 1943. [498] Later, school officials placed Wally's badly wrecked AT6 at the front of the flight line as a constant reminder to approaching cadets as they began their flights. These shock tactics were designed to jolt the young pilots into paying attention to every detail. [499]

As early as 1941, Mrs. Frantie Manbeck Hill began to tend the Miami graves and as more flyers were buried, she often enlisted the help of her family in caring for them. Remembering the fallen heroes was so important to Mrs. Hill that her vigil of care lasted 41 years. [500] During the time that the boys were in training, Mrs. Hill made friends with Cadet Jack Taylor. He was killed in service and his loss was as if she lost her own son. [501] Her stone reads: *Mrs. F.M. Hill of Miami, buried alongside, voluntarily tended these 15 British airmen's graves and helped their loved ones from 1941-1982. These selfless actions were unknown to most.* [502] *She was awarded 'The King's Medal for Service in the Cause of Freedom' by King George VI. Thanks to Mrs. Hill from the graduates of #3 British Flying School's Association.* [503]

Fifteen fallen cadets are buried in the G.A.R. Cemetery north of their school in Miami. *Courtesy Deryk Maddox*

In 1982, after Mrs. Hill was no longer able to continue, Leona DeAtloa Hampton took up the vigil. Later, two sisters, B.B. Wagner and Betty Dimbinski, began to take their place as the caregivers and each year they placed Christmas wreaths on the graves. The Miami Garden Club provided the wreaths decorated by various club women. Dimbinski had served as a USO hostess and knew many of the young chaps well. [504]

Parents of the 15 fallen flyers sent messages to be placed on each man's stone as he was buried so far from home. They were:[505]

• 818163 A.C. 2 RAF (VR) Dennis Mitchell, 20; died August 10, 1943: *"Far from home his country called. Forever in the hearts of his family."*

• 1890382 A.C. 2 Frederick Beverley, 20; died May 30, 1945: *"Your smile, your song, your love will live forever in our hearts."*

• 1628999 A.C.2 Cecil Riddell, 20; died May 2, 1945: *"Our beloved boy. Deeply mourned by all at home. We will remember him."*

• 1810730 A.C. 2 William Mann, 30; died March 21, 1945: *"Rest in Peace."*

• 1572837 L.A.C. James Boyd, 20; died December 14, 1944: *"At rest. Beloved only son of Mr. and Mrs. John Boyd, Prestwick, Scotland."*

• 1807868 A.C. 2 Kenneth Raisbeck, 26; died June 26, 1944: *"Greater love hath no man that he lay down his life for his country."*

• 1159658 L.A.C. Walter Elliott, 29; died October 20, 1943: *"Rest in Peace."*

• 1561238 L.A.C. William Spiers, 20; died April 8, 1943: *"Too soon, too soon comes death to show we love more deeply than we know."*

• 574445 L.A.C. Ralph Price, 19; died October 3, 1942: *"God moves in mysterious ways His wonders to perform."*

• 1338515 L.A.C. Donald Harfield, 17; died August 31, 1942: *"God gave him to us. God has taken him away. Thy will be done."*

• 1320081 L.A.C. Herbert Hacksley, 23; died August 31, 1942: *"Loved and remembered always."*

• 1335301 L.A.C. Harold Burman, 21; died August 31, 1942: *"Live on Brave Boy! Thy spirit still inspires."*

• 1479901 L.A.C. Alan Brown, 19; died August 31, 1942: *"His short life was a life of service."*

• 1199099 L.A.C. Peter McCallum, 19; died November 13, 1941: *"Only son of M.M. and R.E. McCallum of England. Forever in our thoughts."*

• 207274 L.A.C. Fred Tufft, 31; died October 27, 1941: *"He nobly gave life in all his fullness to save us."*

These fifteen young men are true testaments to the ultimate sacrifice some were called upon to make in the defense of their beloved country

Many Miami Cadets Have Led Productive Lives

The contributions many of the men created for themselves in their lives after their training at Miami have been fruitful. Mr. Raymond Baxter, a No. 3 BFTS graduate, became very successful in his native country. His obituary states:

Raymond Baxter, who has died aged 84, was a wartime Spitfire pilot who became one of the BBC television's best-known outside broadcasters, and the first presenter of the popular BBC1 science program, "Tomorrow's World."

Baxter's lean, craggy face and emphatic punched sentences exuded the school-boyish enthusiasm he felt for the up-to-the-minute gadgetry over which he presided for twelve years. Among many "marvels" – one of his favorite descriptions – he introduced to viewers were the electron microscope (1965), the breathalyzer (1967), the pocket calculator (1971), and the barcode reader.

With his American wife Sylvia he demonstrated the first video game, a version of table tennis called Pong, and in another edition he gave the signal for live rounds to be fired at an assistant wearing a new bullet proof vest.

He was, said a later editor of "Tomorrow's World", "the messenger for the great white heat of technological advance which in the 1960s was gong to sort out all our lives". Always preferring the role of "action man", Baxter reported live from the first passenger hovercraft crossing of the Channel in 1966, and made many flights on the Concorde, his first when it was still no more than a prototype fuselage.

He said of the "marvel" of hover flight: "We thought it was going to solve everyone's transport problems. But it just turned into a high-speed ferry on one of the most overcrowded waterways in the world. Its exciting amphibian potential was never exploited."

Baxter parted company with "Tomorrow's World" in 1977, after falling out with its new producer, Michael Blakstad. Blakstad allegedly called the gravel-voiced presenter "the last of the dinosaurs" and Baxter allegedly said "he could never work with someone who rode to the office on a bicycle". He retired hurt and bewildered to the comforting calm of his Queen Anne house and 32-acre garden at Denham in Buckinghamshire.

- A year later he set moved to a more modest home in Henley, where he lived until his death. He set up a business which made commercials for British Leyland and promotional videos and films for business conferences. He also became Honorary Admiral of the Association of Dunkirk Little Ships.

Raymond Frederic Baxter was born on January 25, 1922 at Ilford, Essex, the son of a schoolmaster, and educated at Ilford County High School. As a boy his enthusiasms were divided among motor cars, aeroplanes, and music. He learned the violin and sang as a boy soprano, but at the tender age of 14 he flew with Alan Cobham's flying circus for 10s6d a day (about 25 pounds in today's money). He said later: "I've had a love affair with planes ever since."

At 18 he joined the RAF at the recruiting centre at Lords and disguised as a civilian, was sent with other hopeful pilots to train in Canada and the American Mid-west. He qualified as a pilot (No. 3 BFTS, Miami, Oklahoma) and was posted to No. 65, a Spitfire V fighter squadron based in Scotland, which mostly dealt with shipping reconnaissance.

In the New Year of 1943, following the Operation Torch landings in Algeria and Tunisia, Baxter was posted to the North-West Africa Air Force, joining No. 93, a Spitfire IX squadron covering the First Army, in July. After a year, he was sent home to instruct at No. 61 Operational Training Unit in Shopshire; though he had disliked being taken out of active service, he later said the experience was "just about the happiest six months of my life". Instructing fighter pilots for the last three weeks of their training offered, he recalled, "all the fun of operational flying and the responsibility of leadership without having the enemy around".

In September 1944 he returned to operational flying himself with No. 602. This was a Spitfire IX squadron (re-equipped from November with XVIs) which had just been recalled from Normandy. From Coltishall in Norfolk, he was one of the pilots who dive-bombed Germany's V-2 sites. He was mentioned in the dispatches for his part in these raids.

Towards the end of the war, while still a serving officer, he got his first job in broadcasting, as an announcer with the Forces broadcasting station in Cairo. He then spent two and a half years as deputy director of the British Forces network station in Hamburg. He told an interviewer: "It was a complete course in broadcasting – you had to be able to do everything yourself, and to teach other people at the same time." He was finally demobbed in December 1946 in the rank of flight lieutenant.

Back in Britain the BBC took him on as an outside broadcast commentator, specializing in "mechanical sports" – flying, motorcycling and motor racing. The BBC appointed him their motoring correspondent in 1960 and between 1967 and 1968 he was also director of publicity for the BMC car group. He was part of the BBC commentating team on numerous Royal occasions, including the Coronation and the Queen's State visit to Canada. Baxter also covered the state funerals of Churchill and of King George VI.

Baxter's trademark in covering international sporting events was to double as commentator and competitor. He took part in 12 Monte Carlo rallies, and although his team never won he was twice among the highest placed British competitors. He also took part in the Tulip, Alpine and RAC rallies and in 1953 covered the New Zealand Air Race as a member of the crew of a competing BEA Viscount.

Leaving "Tomorrow's World" marked the end of his regular association with the BBC but he continued to make screen appearances on special occasions. He covered the Farnborough Air Show many times – his commentary during the 1950s was responsible for inspiring many to take an interest in aviation. He was also involved in events to mark the 50th anniversary of the Dunkirk evacuation.

In 1997 and 1998 he returned to his old program for two series of retrospectives, TW: Time Machine, recapturing outstanding moments from the programs of the 1960s and 1970s. In July 2000 he was invited to Earls Court to participate in the presentation of the first Raymond Baxter Award for Science Communication. To his surprise and delight, he discovered that he was also the first recipient.

Baxter married, in 1945, Sylvia Kathryn Johnson, of Boston, Massachusetts; she died in 1996. They had a son and a daughter.

In the 1990s it was revealed that Baxter was, by marriage, the uncle of the avant-garde American sculptor Carl Andre, of Tate bricks fame – or, as modern art buffs know it, Equivalent VIII (1966, now in Tate Modern's permanent collection). Baxter told intrigued interviewers that he had "total respect" for, if not total understanding of, his nephew's work. Visiting Andre in New York, Baxter asked: "So where is this sculpture?" "You're standing on it," Andre told him. [506]

As course after course graduated and returned to fly in the defense of the freedom of their country, these brave young men never hesitated to do what was required of them upon their return to England and to the war that raged about them. Above and beyond what they were ordered to do, many performed exceptional acts of kindness as they went about their duties. Another former cadet, Retired Colonel Gail Halvorsen, Course 18 at Spartan School of Aeronautics in Miami, was well known as the "Candy Bomber." [507] Colonel Halvorsen began dropping packets of candy and chewing gum to the children of Berlin during the Berlin airlift in 1948 and 1949 as he delivered supplies to the area. He saw the children watch each drop and it was not long until he was talking to them and realizing how much they did not have. That is when he decided to hand out the few pieces of chewing gum which he divided into half. The kids did not fight but their eyes got big and they passed around the gum wrapper, sniffing it for its wonderful smell. It was then that he promised to return the next day with enough gum for each child to have a whole stick. The kids asked for a sign so that they would recognize his plane and that is when he told him that he would wiggle his wings as he approached. From then on he was called "Uncle Wobbly Wings" and "Der Scholkoladen Flieger (The Chocolate Pilot)." He continued to bring in rations saved for the children by his buddies and eventually, the news of what he was doing spread. Finally, his general in command gave him official permission to continue the project and he, along with his buddies, dropped handkerchiefs for parachutes. The children would send them back through the mail for refills and along with the handkerchiefs came heartwarming letters of gratitude.

Years later, Colonel Halvorsen was approached by one of the children from so long ago. Her letter was one of the ones that he had kept for so many years and he was amazed when she found him and introduced herself. Although he had long since decided that the people of Berlin were the heroes, it was gratifying for him to learn that she and others had remembered him and what he had done to help them endure a very hard life.[508] Colonel Halvorsen wrote a book about his experiences and as late as 2000 he took part in an effort to deliver much needed goods such as clothing, food, and toys to isolated Pacific islanders.[509]

Cadets from Miami marched before large crowds of friends in wings parades.

Courtesy Northeastern Oklahoma A&M College

1945 Sees the Closing of Spartan School of Aeronautics

Spartan School of Aeronautics was purchased by the Defense Plant Corporation some time before August, 1944, as was the standard practice for all the Army Air Force's contract flying schools.[510] In Miami, the August 23, 1945 Wings Parade was the last for Spartan School of Aeronautics.[511] As a band from Camp Crowder played a series of stirring marches, the class of 25 Course marched in review. Saddened by the closing of the school and their return to their homeland, they knew that it was the last time many would see their friends in Oklahoma again. In their hearts, they knew they also had ended an era in which 25 successful training courses left capable and well trained pilots who had bravely gone home to serve their land. Course 25 was just as proud to return to "guard the peace brought at the high cost of war."[512] Two other courses, 26 and 27, never completed their training in Miami due to the end of the war and, as a result are not officially listed in association rosters.[513]

TOP: Members of a graduating class at No. 3 BFTS stand at attention as their classmates receive their coveted wings. This was the highlight of their training in America and although most were saddened to leave their new friends in America, the proud young men were eager to return to their homes and to fight the Nazis. *Courtesy Northeastern Oklahoma A&M College*

INSET: Deryk Maddox is presented with his wings. *Courtesy Deryk Maddox*

Both RAF and American cadets paraded along the flight line tarmac. Aircraft were lined up in a precise fashion as the parades would pass in review, turn, and then form again in front of the reviewing stand. Officials and guests would gather at this location while they listened to inspiring speeches given by the RAF Wing Commander and by the USAAF representative. The Yanks received silver wings as well as RAF wings. The American wings were placed on the left side while the RAF wings were placed on the right side of his coat front. It was a proud and humbling moment for each man who stepped forward at his graduation as he realized that far away his family, friends, and country had depended on his struggle for achievement despite the many sacrifices which came with his study in a country half way around the world from his home.

Just hours after the presentation of their wings, graduation parties were held and Arkansas White Lightning flowed freely. One course held their graduation party at the Connor Hotel in Joplin, Missouri.[514] Although Oklahoma was a dry state, generous citizens of the town who were invited to the parties often produced the alcohol. Most of these parties were stag so there was no dancing but the men, their instructors, and guests enjoyed delicious steaks and all that went with them. While they were being served, talking and singing erupted all around the room. The men were allowed to sleep late the following morning but only hours later the new RAF officers were placed on buses and taken to the train station, where they left on their long journey back to England and the war. Their American counterparts missed them immediately, but were soon busy settling into their own routines. These Americans were discharged as Aviation Cadets and then commissioned as 2nd Lieutenants. It was then that they received their orders to aerial duty as pilots.

Newly graduated pilots from Miami expressed their sentiments when they listed a few of the things they would never forget. They mentioned their first sight of Miami, their first solos, the weekends in surrounding towns such as Parsons, Kansas, Joplin, Missouri, and Tulsa, Oklahoma; their desire to bomb the Link Trainer department; mail call; and the friends they made during their stay in America. All were very appreciative of the kindness they received from the staff, instructors, and the people of Miami.[515]

A group of new No. 3 BFTS graduates pose during their graduation dinner. *Courtesy Ponca City Library*

In the summer of 2005, No. 3 BFTS Association officially disbanded at their final reunion in Miami, Oklahoma. Later in the year, former chairman Deryk Maddox recalled that the group had joined together for their first reunion in 1948 at a dinner in London. The reunion was chaired by Wing Commander Roxburgh and was organized by the late Hugh Hallet. The reunion again was held in 1949 with sporadic reunions through 2005.

In 1982, George Mayer, a former instructor and Miami businessman, and Keith Leigh organized the first return visit to Miami. The occasion was so well received that the following year several American friends journeyed across the pond and were entertained in the homes of some of the former English cadets. Five years later, in 1987, another group of English members journeyed to Texas where they enjoyed the Confederate Air Force air show. They returned to Miami for another visit full of fond memories of time spent in the Oklahoma town. To celebrate the 50th anniversary of the school, the group met in Miami in 1991. Ultimately, there were four United States reunions and eight in the United Kingdom before, finally, for a variety of reasons, the association decided to disband in 2005. Over a period of twenty years, the group maintained either an *Open Post* magazine or newsletter; hosted an internet website; provided a comprehensive historical archive of their flying school at the Miami college; donated memorabilia and photographs to the Dobson Museum and the Kirkpatrick Museum; and established a Memorial Award Fund at Northeastern Oklahoma A&M College in Miami, which helps young people wishing to further their education.

As the final edition of the *Open Post* closed, the efforts of No. 3 BFTS remained active until in November, 2005, the final collection of No. 3BFTS memorabilia was shipped to the British Embassy in Washington, D.C., courtesy of the Royal Air Force. It was later transported to Miami where it resides in the RAF archives housed in the library of Northeastern Oklahoma A&M College.

Miami cadets gather at the rail station in Miami as they prepare to depart for Moncton, New Brunswick on their way back to England. Many sweethearts dressed in their Sunday best as they sent their English pilots off to war. *Courtesy: Kay Wessale Schnieber*

Proudly standing near an American flag at a Miami, Oklahoma home, young British pilots represent the democratic freedom that United States citizens still enjoy today because of their dedication to peacekeeping as they prepared in Oklahoma. *Courtesy Northeastern Oklahoma A&M College*

Mounds of kit bags are piled and ready for loading onto the trains which would carry their owners back to Canada. *Courtesy Ponca City Library*

Miami cadets gather their luggage and souvenirs while they waited for the train to Moncton, New Brunswick. For many years later and up until 2005, a dedicated group of these men returned for reunions in Miami. *Courtesy Ponca City Library*

The last of the Spartan School of Aeronautics buildings as they stood in December, 2004. By the summer of 2005 these buildings had met the fate of a bulldozer. *Courtesy Paula Denson*

Epilogue

And so the magical journey to Oklahoma and the happy and somber experiences which visited them along the way quickly came to an end for the English men graduated as pilots from Darr School of Aeronautics in Ponca City and the Spartan School of Aeronautics at Miami, Oklahoma. From the beginning of their journey to their eventual return to England, these young men found a quality of life of which they could have only dreamed of back home in war-torn England. Their awe of the vastness of America and the friendly nature of everyone they met as they traveled to the heartland of the United States became part of the epic resulting in the winning of World War II.

Distant clouds, rising to lofty heights on the horizon, silhouetted a lone Harvard AT6 as quiet came to the fields where the British trained in Ponca City and Miami. *Courtesy Ponca City Library*

Geoff Kirrage and Tommy Tucker pose informally during their stay in Ponca City.
Courtesy Phyllis Mann Wiles

APPENDIX

Every effort has been made to list all known cadets, staff members, and other personnel associated at both schools, using what limited resources were available. However, no claim has been made as to the accuracy of these rosters. It is highly logical that there are many names which yet remain to be discovered.

If you have additions, corrections, or wish to share the full names of a person or persons listed in this appendix, please contact the author at **paula@pkdenson.com.**

This unknown American cadet at Miami proudly stands on the wing of his aircraft in preparation for aerial maneuvers.

Courtesy Northeastern Oklahoma A&M College

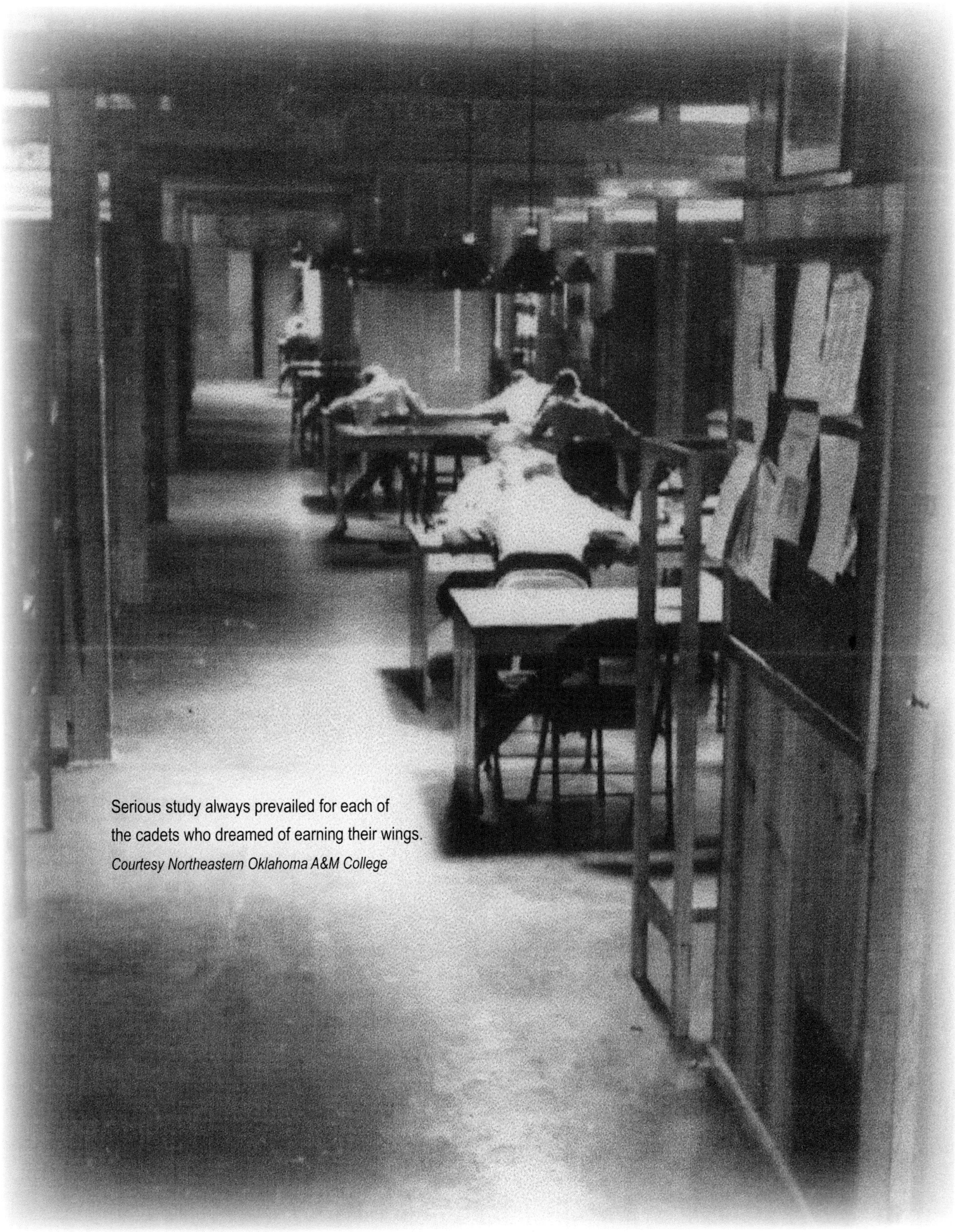

Serious study always prevailed for each of the cadets who dreamed of earning their wings.
Courtesy Northeastern Oklahoma A&M College

— DARR SCHOOL OF AERONAUTICS · PONCA CITY, OKLAHOMA —

LAST NAME	FIRST NAME	RANK	COURSE NO.
Aasen	John C. "Jack"	Flying Instructor	0
Ablett	E.T.	LAC	08
Adams	Alfred "Joe"	LAC	02
Adams	"Andy"	LAC	05
Adams	John	LAC	04
Addington	Gurth L.F.	LAC	06
Addison	Douglas	LAC	04
Adkins	Emmet E.	Air Maintenance	0
Adamec	Robert Francis	Air Maintenance	0
Aedy	Kenneth G.	LAC	13
Ahlstrom	R.S.	Flying Instructor	0
Aikman	William L.	Cadet W/Cdr.	16
Ainsworth	"Stan"	LAC	07
Alair	Henry C.	Radio Department	0
Albins	C.W. "Cedric"	LAC	13
Albrecht	Hardie Martin	AV/C; USAAF	15
Alden	Vincent Earl	Unknown Position	0
Alderman	Frederic C.	LAC	13
Alexander	Joe	LAC	03
Allam	John W. or A.J.	P/O	15
Allen	Christopher Oswald Colquhoun "Gubby"	LAC	10
Allen	K.F.	LAC	17
Allison	C.K.	LAC	17
Allwood	P.	LAC	11
Anderson	Harry H.	Dispatcher	0
Anderson	John W.	LAC	15
Anderson	Paul W.	Flying Instructor	0
Anderson	Robert "Andy"	LAC	15
Anderson	W.D.	LAC	14
Andrews	T.N.	LAC	17
Anscomb	Robert "Bob" "Moon"	LAC	10
Armstrong	"Bob"	LAC	10
Armstrong	"Paddy"	LAC	01
Arney		AV/C; USAAF	17
Arnold-Forster	Nigel M.	LAC	15
Arthur	Paul	Air Maintenance	0
Artis	Bruce F.	Flying Instructor	0
Ashford	D.M. "Dave"	LAC	04
Ashman	Lewis	LAC	17
Ashton	M.L. "Dixon", "Dixie" or "Dick"	LAC	08
Aspinall	Norman	LAC	06
Aston	Robert A. "Bob"	P/O	11
Atkinson		LAC	11
Atkinson	J.	LAC	14
Auld	J. "Tick"	LAC	05
Austin	W.	LAC	06
Aykroyd		LAC	03
B	J.	LAC	07
Bacher	Marion L.	Air Maintenance	0
Backett	Harold	LAC	0
Bailey	Cyril	LAC	05
Bailey	L.A.	Fl. Sgt.	0
Baillie	Edward McAlister	LAC	15
Baillie	Lewis H.	Office Manager; Staff	0
Bainbridge	J.	LAC	16
Baker	J.P.	LAC	11
Baker	"Tex"	LAC	07
Baker	W.H.	AV/C; USAAF	17
Ball	Barbara	Air Maintenance	0
Ball	C.A.	Wing Commander	0
Bames	W.H.	LAC	02
Banks	Carl O.	Dispatcher	0
Banks	G.W.	LAC	17
Banks	Glen L.	Dispatcher	0
Banton	David I.	LAC	11
Barber		LAC	12
Barbre	W.A. "Bill"	AV/C; USAAF	06
Barker	E.J.M.	Fl. Lt.	0
Barnard	J.	LAC	13
Barnes	F.A.	LAC	17
Barnes	F.J. "Freddie"	LAC	01
Barnes	Harold	Office Staff	0
Barnett	Elmer H.	Air Maintenance	0
Barnwell	Robert "Bob"	LAC	13
Barr	Clarence E.	Flying Instructor	0
Barr	"Duncan"	LAC	04/05
Barrett	E. Malcom "Mal"	LAC	09
Barrett	"Francis"	LAC	05
Barrington	John S.	LAC	07
Barritt	H.M. "Eddie"	LAC	01
Barry	J.C.	AV/C; USAAF	16
Barton	K.	LAC	11
Bashford	Roy Stanley	LAC	15
Bass	Robert Leon	Unknown Position	0
Baswell	J.	LAC	06
Battarbee	C.	LAC	04
Batten	H.J.	LAC	15
Batten	R.J.	LAC	17
Baumber	Charles L.	LAC	13
Baxter	Luther T.	Air Maintenance	0
Bayley	R.M.	LAC	09
Baylis	J.F.H.	LAC	13
Beal	Alan	LAC	02
Beaman	D.T.	2nd Lt.; USAAF	0
Beauchamp	J.E.	LAC	05
Beaumont	E.G.	LAC	10
Beavis	A.S. John	LAC	07
Beckett	"Joe"	LAC	04
Beech		LAC	12
Beeching	E.	LAC	12
Beindorf	J.A.	Flying Instructor	0
Bell	H.	LAC	14
Bellamy	K.J.	LAC	08
Bengree	Hugh C.F. "Taffy"	LAC	12
Bennett	D.J. "Ginger"	LAC	03
Bennett	J.	Sgt.	12
Benson	W.A.	LAC	12
Bettles	Kevin J.	LAC	08
Beveridge	R.N. "Haggis"	Sqd. Ldr.	09
Bibby	Ed	LAC	04
Billiam	John	LAC	0
Bince	W.L.H.	LAC	16/17
Bishop	Rufus R.	Air Maintenance	0
Bissell	R.H.	LAC	15
Bjorn		Lt.; USAAF	0
Black	J.D.	Air Maintenance	0
Blackett	H.E.	LAC	12
Blair	Don	LAC	04
Blair	J.	LAC	08
Bleakley	Irene	Office Staff	0
Bluck	J. "Johnny"	LAC	01
Blue	Frank C.	Air Maintenance	0
Blue	Homer D.	Air Maintenance	0
Blythe	R. "Bob"	LAC	04
Boden	G.A. "Ashley"	Sgt.	14
Boilla	Romi	LAC	0
Boivin	L.C. "Les"	LAC	07
Boland	C.E.	AV/C; USAAF	15
Bolick	Jack N.	Link Trainer Instructor	0
Bolton	R.A. "Dick"	LAC	09
Bolton	G.F.	LAC	11
Bolton	S.	LAC	08
Bolton	Albert W.	AV/C; USAAF	17
Boniface	G.J.	LAC	10
Boon	J.F.	LAC	17
Booth	J. Simon	LAC	13
Boren	Roger S.	Air Maintenance	0
Bormet	George	Flying Instructor	0
Bourne	A.G.	LAC	17
Bowers		AV/C; USAAF	16
Bowhay	Gene W.	Parachute Staff	0
Bowhay	A.J.	Air Maintenance	0
Bowhay	G.A.	Air Maintenance	0
Bowles (Taylor)	Lillian	Link Trainer Instructor	0
Bowman	Ben F.	Watchman	0
Bowman	Glenn F.	Watchman	0
Bowman	J.H.	Flying Instructor	0
Boyd(e)	E.	LAC	11
Boyle	Charles "Robbie"	LAC	09
Boys	Peter K. E.	LAC	01
Bradford	W.	LAC	11
Bradshaw	Reg	LAC	02
Brady	S.C.	Capt.; USAAF	0
Bray	Geoffrey	LAC	15
Breakwell	T.	LAC	17
Brebner	J.L.	LAC	16
Brewer	David G.	LAC	14
Brewer	R. J. "Bob"	LAC	01
Bridges	A.M.	AV/C; USAAF	14
Brobyn	R.G.	LAC	17
Brockbank	C.A.V.	LAC	17
Brockington	C.B. "Ben"	LAC	04
Brodie	"Alan"	LAC	05
Bromley	Hayes	LAC	0
Brooks		LAC	11
Broomfield		LAC	13
Brown		LAC	17
Brown	Bertie	LAC	01/02
Brown	"Billie"	LAC	12
Brown	E.J.	LAC	15/16
Brown	L.H.	LAC	09
Brown	Leonard F.	Mess Hall	0
Brown	Lewis E.	Radio Department	0
Brown	V.	LAC	11
Bruce-Gardner	Brian C.	LAC	15
Brush	O.C. "Johnny"	LAC	08
Bryant	Mack S.	Air Maintenance	0
Buchanan		LAC	06
Buchanan	J.J.	Corp.	0
Buchanan	"Sandy"	LAC	13
Buecchler	P.W.	AV/C; USAAF	17
Bull	G.G.	LAC	17
Bunker	J.N.	LAC	14
Burham	E.K.	AV/C; USAAF	13
Burk	Alexander B.	Watchman	0
Burke	Ocie Harold	Air Maintenance	0
Burnett		LAC	03
Burnett	Ronald "Ron"	LAC	01
Burnett	Weldon F.	Air Maintenance	0
Burnup	John	LAC	08
Burr	J.T. "Jack"	LAC	01/02
Burrell	L.H.	LAC	16
Burrell	R. "Bob"	LAC	04/05
Burright	Weslely H.	Air Maintenance	0
Burson	B.	LAC	12
Burtenshaw	D.F.	LAC	09
Burton	Benjamin	LAC	13
Burton	H.	LAC	16
Bush	E.	AV/C; USAAF	11
Butcher	Leslie	LAC	15
Butler		LAC	15
Butler	I.E.S.	LAC	17
Byrne	Douglas N.	P/O	15
Cable	C.G.	LAC	17
Cadman	R.C.	LAC	09
Calder	I.R.F.	LAC	14
Callister	Grant	AV/C; USAAF	16
Cameron	"Gregor"	LAC	07
Cameron	J.M.	LAC	10
Campbell		LAC	13
Campbell	Jim/Jock	LAC	04
Campbell	N.P.	AV/C; USAAF	15
Campbell	S.W.	LAC	11
Campbell	Thomas	Flying Instructor	0

LAST NAME	FIRST NAME	RANK	COURSE NO.
Cann	L.O.	LAC	13
Capp	Paul E.	Air Maintenance	0
Carder	L.	LAC	12
Carey	G.W.	LAC	08
Carlson	Henry	USAAF	14
Carpenter	R.H.	Flying Instructor	0
Carter	J.N.	LAC	17
Carter	K.	LAC	13
Carter	R.C.	LAC	13
Carter	Ray Eugene	Unknown Position	0
Carver	D.	LAC	17
Casey	M.E.	LAC	17
Cassidy	J.V.	LAC	14
Castle	Ivan	LAC	10
Castle	J.N.	LAC	11
Cathcart	James "Jim"	LAC	16
Catlin	D.W.	LAC	14
Caudill	C.	AV/C; USAAF	13
Cavnar	Bob J.	AV/C; USAAF	17
Cawthorn(e)	E. "Eddie"	LAC	10
Chadwick	John B.	LAC	11
Chambers	Leonard C.	Air Maintenance	0
Chambers	P. "Jerry"	LAC	05
Chandler	John R.	LAC	02
Chapin	John O.	Flying Instructor	0
Chapman	D.A.	LAC	17
Chapman	Merle	Air Maintenance	0
Charlesworth	A.	LAC	17
Charman	Ken P.	LAC	15
Chattin	Peter W. "Pete"	LAC	05
Childers	Floyd A.	AV/C; USAAF	13
Chokola	P.F.	AV/C; USAAF	13
Christie	F.W.	LAC	14
Chronister	T.F.	Flying Instructor	0
Clampitt	Belle	Canteen	0
Clapham	Kenneth	Flying Instructor	0
Clark	F.S.M.	LAC	10
Clark	Noel R.	LAC	02
Clark	R.	AV/C; USAAF	11
Clarke	Colin Reed	LAC	15
Clarke	John H.	LAC	16
Claxton	Derrick W. "Clac"	LAC	03
Cleaver	L.	LAC	11
Clee	Roy B.	LAC	15
Cleland	"Bill"	LAC	09
Clement	George W.	LAC	02
Clements	J.D.	LAC	14
Cliff	G.D.M. "Geoff"	LAC	01
Clifton	George	LAC	13
Clinton	Charles	Mess Hall	0
Clutter	Elmer H.	Flying Instructor	0
Clyne	J.	LAC	05
Coade	S.G.	LAC	10
Coale	C.W.	AV/C; USAAF	13
Coates	C.R.	LAC	15
Cockcroft	G.N.	LAC	04
Coe	Dewey E.	Air Maintenance	0
Coggan	P.	LAC	11
Cogman	Harold "Coggie"	LAC	01
Cohen	"Dan"	LAC	13
Coldham	Peter "Tony"	LAC	15
Cole		LAC	07
Coleman	Francis G.	Ground School Staff	0
Colley	G.E.	LAC	10
Collier	R.E.	LAC	10
Collins	E. Hugh	LAC	02
Collins	M. "Mike"	LAC	01
Conrad	Harry C.	Unknown Position	0
Cook	A.D. "Art"	LAC	07
Cook	Edward Ted	LAC	15
Cooney	H.F.	LAC	11
Cooper	N.E.	LAC	10
Cooper	R.D.	Flying Instructor	0
Cooprider	T.B.	Air Maintenance	0
Copley	Isaiah	Radio Department	0
Corewyn	W. G "Bill"	LAC	02
Corlett	A. "Jack"	LAC	01
Corzine	Lee	Flying Instructor	0
Coubrough	A.M.	LAC	14
Coulthard	J.T.	LAC	10
Courson	Arthur T. "Bud"	Flying Instructor	0
Courtman-Stock	Norman W.	LAC	15
Coveney	Alfred R.	LAC	12/13
Cowell	Keith H.R.	LAC	15
Cowley	N.C.	LAC	13
Cox	"Mike"	LAC	12
Crawley	J.S. "Ned"	LAC	12
Creighton	Jack	LAC	12
Crierie	"Jimmy"	LAC	16
Crisp	D.H.	LAC	17
Crosby	R.H. "Bing"	LAC	10
Croshaw	J.G.	LAC	14
Crossley	K.O.	LAC	12
Crouch	Bobby R.	Air Maintenance	0
Crouch	Dennis W.	LAC	03
Crow	S.G.	LAC	02
Crowley	E.	Wing/Cdr., VC	02
Crowther		Fl. Sgt.	0
Ctummett	Earl	Unknown Position	0
Cubey	J.K. "Ken"	LAC	09
Cumming		LAC	11
Cummings		LAC	12
Cunliffe	Ronald A.	LAC	15
Cunning	Charles "Chuck"	Flying Instructor	0
Cupp	G.W.	Unknown Position	0
Curry	J. James	LAC	12
Curtis	Dave	LAC	12
Curwen	Leslie Wilkinson	LAC	15
Cust	P.G.	LAC	05
Custer	E.D.	Air Maintenance	0
Cusworth	Frank H.	LAC	15
D(C)yphers	E.L.	AV/C; USAAF	13
Dale		LAC	13
Dalgleish	"Met"	LAC	04
Dalrymple	H.P.	Flying Instructor	0
Dalton	F.O.	AV/C; USAAF	11
Dalton	J.F.M. Jack Fred Peter	LAC	12/13
Daly	G.V.	LAC	11
Dame	Earl F.	Air Maintenance	0
Dane	Paul	LAC	13
Daniel	Jack B.	Air Maintenance	0
Daniels	Peter E.	LAC	02
Daniels	T. Brian	LAC	15
Darm(n)sell	W.A. "Tony"	LAC	01
Darr	Harold S. "Hal"	Darr CEO	0
Dashwood	Roland P.	LAC	06
Daum	William R.	AV/C; USAAF	14
Davidson	John W.	LAC	14
Davies		LAC	11
Davies		LAC	16
Davies	H.R. "Howard"	LAC	13
Davies	Owen L.	LAC	06
Davies	R.S. "Taffy"	LAC	07
Davies	William H. "Bill"	Sgt.	15
Davis	"Dickey"	LAC	07
Davis	H.D.	LAC	11
Davison	R.A. "John"	LAC	09
Davy		LAC	04
Davy		LAC	11
Dawson	T. P. (C.) Thomas	LAC	02
Day		LAC	11
Day	"Nancy"	LAC	09
De La Rue	S. "Max"	LAC	15
Deans	James L. "Jim"	LAC	05
Dear	Norman E.C.	LAC	05
DeClaspel	J.F.	Flying Instructor	0
Deda	George H.	Mess Hall	0
Degan		LAC	05
Delahaye	P.F.	LAC	10
Dell	R.C.M.	LAC	09
Denman	Gerald	Flying Instructor	0
Denton	E.W.	LAC	17
Derveloy	Thomas	Office Staff	0
DeVoll	R.	AV/C; USAAF	16
DeWitt	Bernard G.	Flying Instructor	0
Dick	Alec W.M.	LAC	04
Dickie	K.J.A. "Ken"	LAC	02
Dickson	D.	LAC	17
Dinnage	Harry A.	LAC	15
Dinwood		LAC	05
Dirham	Karl M.	Air Maintenance	0
Dirham	Walter	Air Maintenance	0
Ditchfield	R.H.	LAC	13
Dixon	"Dickie"	LAC	05
Dobson	M.F.M.	LAC	17
Dodd	Irma	Mess Hall	0
Dodson	R.R.	AV/C; USAAF	14
Dohrman	David S.	AV/C; USAAF	14
Dolphin	H.A.	LAC	16/17
Dolphin	T.	LAC	02
Donohoe	C.A.	AV/C; USAAF	17
Doorey	Stan G.	LAC	02
Dorman	E. Jack	LAC	02
Dormer	Reginald "Reggie"	LAC	15
Douglass	A. Graham "Copper"	LAC	10
Downs	J.T.	LAC	17
Dowthwaite	Peter W.	LAC	17
Drew	R.J. or A.E.	LAC	17
Driscoll	R.M.	Air Maintenance	0
Driver	"Johnny"	LAC	13
Drury	T.	LAC	11
duBoulay	R.W.H. "Roger"	LAC	07
Duckworth		LAC	13
Duguid	Eric	LAC	17
Duker	Vernon	LAC	17
Duncan	M.M.	LAC	12
Duncliffe	Derek A.	LAC	09
Dunford	J.R.	Sgt.	12
Dunworth	Alan B.	LAC	14/15
Durbridge	Keith	LAC	03
Durlin		AV/C; USAAF	16
Dutton	Howard	Flying Instructor	0
Dux	Godfried J.	Sq/Ldr; Flying Instructor	0
Dwyer	A.J. or C.J.	LAC	14
Dye	R.M.	LAC	15
Dyke	"Bill"	LAC	07
Eadie	Roy A.	LAC	05
East	William T.	Air Maintenance	0
Easton-Smith	D.	Fl. Lt.; Staff	0
Eaton	S.J.	LAC	06
Eddy		LAC	03
Ede		LAC	16
Edison	Robert "Bob"	Flying Instructor	0
Edmunds	M.S.	LAC	06
Edmunson	William H.	LAC	02
Edwards	D.W.	LAC	17
Edwards	Helen Louise	RAF Secretary	0
Edwards	George	LAC	03
Edwards	Norman	LAC	15
Eeds	Holt	Air Maintenance	0
Eggleston		LAC	12
Eglington	George	LAC	01/2
Elkins	A. George	Flying Instructor	0
Elkins	James W.	Link Trainer Instructor	0
Eller	Vernon E.	AV/C; USAAF	14
Elliott	G.H.	LAC	16
Ellis	DRoss	USAAC Flight Director	0
Ellwood	Richard S.	LAC	15
Elms	Frank	LAC	08
Elsden		LAC	15
Emery	"Frank"	LAC	07
England	B.A.	LAC	10
English	Ken	LAC	02
Enos		LAC	16
Essen	Malcom Frederick	Sgt.	15

LAST NAME	FIRST NAME	RANK	COURSE NO.
Estes	George	Mess Hall	0
Estes	Lemuel E.	Janitor	0
Etheridge	"Jimmy"	LAC	13
Etherington	Harry M.	LAC	06
Euston	Vincent George	LAC	15
Evans	H.K.	LAC	04
Evans	H.T.	LAC	09
Evans	Loyd E.	Radio Department	0
Evans	R.I. "Bob"	LAC	03
Evans	R.P.	LAC	12
Evans	Thomas O.	Air Maintenance	0
Everard	R.C. "Russell"	LAC	13
Evinger	Jack H.	Flight Dispatcher	0
Ewing	"Jim"	LAC	13
Eyres	R.D.W. "Jumbo"	LAC	17
F	G.A.F.	LAC	13
Fabian	L.	LAC	13
Farrish	F.	LAC	11
Farmer	E.	LAC	14
Farnam	R. Max	Air Maintenance	0
Farris	R.	AV/C; USAAF	16
Farrow	C.	AV/C; USAAF	16
Farrow	Kenneth	LAC	15
Fasnacht		LAC	17
Faun	R.C.	LAC	12
Favor		LAC	12
Fawcett	Walter Ernest D.	LAC	15
Fayne	J.	LAC	11
Feagin	Marion L.	Watchman	0
Fear	F.	LAC	16
Feather	C.	LAC	15
Fedler	William F.	Air Maintenance	0
Fell	M. or H.	LAC	14
Fellers	Clyde W.	Flying Instructor	0
Ferguson	William "Bill"	LAC	13
Fern	Brian E.	LAC	09
Fielding-Johnson	Hugh H.	LAC	06
Flannery	John P.	LAC	10
Flenniken	W.S.	LAC	10
Fletcher	Donald James	LAC	15
Fletcher	John "Dennis"	LAC	03
Flood	M.T. "Mike"	LAC	06
Florentine		AV/C; USAAF	16
Flux		LAC	17
Fogden	"Bill"	LAC	11/12
Fogg	Ronald S.V. "Ron"	LAC	15
Formby	G.	LAC	12
Forrest	J.W.	LAC	17
Forster	M.	LAC	11
Foster	D.	LAC	13
Foster	Donavan J. "Don"	LAC	01
Foster	G.A.	LAC	08
Foster	Ralph C.	Air Maintenance	0
Foster	V.R.	Flying Instructor	0
Fothergill	W.B. "Joe"	LAC	10
Fowell	George	LAC	10
Fox	W.	LAC	16
Foxell	J.S.	LAC	17
Fraccaro	Joseph Mario	Unknown Position	0
Francis	L.	AV/C; USAAF	11
Franklin	F. Ernest William Cecil George	LAC	01
Franklin	Warren G.	Flight Dispatcher	0
Frankowitz	J.F.	Flying Instructor	0
Franks	Claude E.	Air Maintenance	0
Fraser	J.	LAC	15
Frayne		LAC	11
Freeman	L.D. "Danny"	Sgt.; Staff	0
Freeman	J. Dee	Air Maintenance	0
Freeman	J.S.	LAC	17
Freer		LAC	12
French	Gordon	Office Staff	0
Freye	L.H.	Flying Instructor	0
Frizzo	Jay	Flying Instructor	0
Fry	R.L.	LAC	16
Fulton	"Mac"	LAC	04
Furness	G.F. "Prof"	LAC	07
Fursman	E.F. "Ted"	LAC	12
G	H.	LAC	10
G	P	LAC	04
Gabrain	R.W.	LAC	13
Gage	H.A.J.	LAC	15
Gage	Margaret	RAF Secretary	0
Galbraith	Donald S.	Flying Instructor	0
Galley	I.S. "Ivor"	LAC	07
Gallivan	P.J.	LAC	11
Galloway	Charles R.	Flying Instructor	0
Galloway	M.	Flying Instructor	0
Game	D.	LAC	13
Garforth	R. "Bob"	LAC	04
Garland	J.	LAC	08
Garnick	Frank Phillip	AV/C; USAAF	15
Garrard	Leslie A.	LAC	11
Garside	Ken	LAC	04/05
Garvey	J.C. "Chris"	LAC	13
Gass	J.R. "Ron"	LAC	15
Gates	John Wesley	Unknown Position	0
Gates	P.R.	LAC	17
Gates	W.	LAC	16
Gee	Francis O.	Mess Hall	0
Gee	J.	LAC	14
Geraghty		LAC	09
Germinaro	Frank	Air Maintenance	0
Gibson	G.W. "George"	LAC	13
Gibson	L.L.	LAC	16
Gilbert		LAC	04
Gilbert	Howard O.	Air Maintenance	0
Gilbert	R.S.	LAC	14
Gill	H. "Jimmy"	LAC	08
Gillespy	Charles F.	Radio Department	0
Gilliam	Claude G.	S.A.A.D.	0
Gilmour	J.C. "Max"	LAC	14
Ginsberg	H. "Bertie"	LAC	17
Glaser	Gerhart Max	Air Maintenance	0
Glaser	"Herb"	Air Maintenance	0
Goddard	Victor F.	LAC	03
Goldman	Garnett	Air Maintenance	0
Golliver	James D.	Unknown Position	0
Gooch	Freddie	LAC	11
Gooch	Woodrow W.	Air Maintenance	0
Goode	R.G. "Pat"	LAC	04
Gooding	L.A. "Len"	LAC	12
Goodman	Edward C. "Ernie"	Air Maintenance	0
Goodship	Wilfred "Venus"	LAC	03
Goodson	Leslie R.	Air Maintenance	0
Gosling	J.E.R. "Jimmy"	LAC	13
Gosting	Donald L.	Air Maintenance	0
Gottschalk	Chester	Air Maintenance	0
Gough	Geof J. "Tramp"	LAC	14
Gough	R.C.	LAC	17
Graf	J. John	LAC	12
Graham	C.E.	LAC	05
Graham	Donald T.	LAC	01
Grant	M.	LAC	14
Grant	R. "Ron"	LAC	06
Graves	Jack W.	Air Maintenance	0
Gray		LAC	06
Gray	G.	LAC	01
Gray	G.	LAC	16
Gray	H.	Sgt.; Staff	0
Gray	"Red"	AV/C; USAAF	13
Gray	T.G.	LAC	17
Greaves	P.K.	LAC	10
Green	B.M.	LAC	11
Green	"Frankie"	LAC	13
Green	"Freddie"	LAC	06
Green	H. "Fred"	LAC	10
Green	John Geoffrey	LAC	15
Green	L.A.	Fl./Sgt.	12
Green	L.A. "Red" "Len"	LAC	09
Greenburgh	Louis "Lou"	LAC	06
Greene	F.A.	LAC	06
Greenway		LAC	12
Greenwood	W.L.H. "William"	LAC	12
Greer	Louis R.	Canteen	0
Gregory	"Albert"	LAC	13
Grey	G.E.A.	P/O	11
Grey	W.	LAC	02
Griffiths	N.E.	LAC	16
Griffiths	W.J.	LAC	14
Grime	A.T.	Fl. Lt.	Staff
Grossoehme	George	Flying Instructor	0
Grover	K.	LAC	11
Grynkiewicz	Terofil P. "Grynki"	LAC	01
Gunn	Horton F.	Flying Instructor	0
Gunnell	P.	LAC	06
Gustafson	H.E.	Air Maintenance	0
Guthrie		LAC	16
Gutshall	William Gibbons	AV/C; USAAF	15
H	G.K.	LAC	11
Haaz	Daniel David	Air Maintenance	0
Haberman	J.	AV/C; USAAF	12
Hadden	William "Bill"	LAC	12
Hadley	A.N.	LAC	03
Halbrook	Arthur J.	Flying Instructor	0
Hale(s)	G.	LAC	11
Haley	Ned	LAC	03
Hall	Bernard	LAC	01
Hall	S.A.	LAC	15
Hall	H. M.	AV/C; USAAF	11
Hallmark	Hayes	Air Maintenance	0
Hallmark	Lloyd	Air Maintenance	0
Hammerton	Derek	LAC	17
Hancock	A.	LAC	16
Hand	John L.	Air Maintenance	0
Handyside	Ian G.	P/O	07
Hanes	Nathaniel	Flying Instructor	0
Hankin	G.	LAC	05
Hanney	Ken	LAC	12
Hanson	E.D.	AV/C; USAAF	12
Haralson	Howard	Air Maintenance	0
Harbridge	Ken B.	LAC	15
Harder	F.	LAC	11
Harding	Tex W.	Parachute Department	0
Hardy	R.	LAC	11
Harkness	Russell	Air Maintenance	0
Harley	E.V. "Vin"	LAC	09
Harr	Reed S.	AV/C; USAAF	17
Harris	A.A.	LAC	09
Harris	A.J.	AV/C; USAAF	11
Harris	R.D.	Air Maintenance	0
Harrison	Len	LAC	13
Harrison	R.O.	LAC	11
Harrole	H.B.	Air Maintenance	0
Hart	Aubrey A.	AV/C; USAAF	17
Harvey		Flying Instructor	0
Harvey	D.W.	LAC	11
Harvey	R.H.G.	LAC	11
Haslam	Raymond	LAC	05
Hastings	Jack	LAC	14
Hasz	Daniel D.	Air Maintenance	0
Hatchard	Vincent	LAC	04
Hathaway	E. John	Staff	13
Hawkins	A.	LAC	11
Hawkins	S.	LAC	16
Hawley	D. "Don"	LAC	06
Hawthorn	R.V.	LAC	17
Hayes	L.E.	Air Maintenance	0
Haynes	B.G.W.	LAC	12
Haynes	G.R.V. "Bill"	LAC	13/14
Haynes	T.	LAC	08
Hazell	G.	LAC	16
Heeley	R. "Red"	LAC	08
Helm	Lloyd G.	Unknown Position	0
Henderson	Bert S.	Air Maintenance	0

LAST NAME	FIRST NAME	RANK	COURSE NO.
Hennessey	P.	LAC	10
Hennessy	Thomas "Paddy"	LAC	01
Hennings	John	LAC	12
Henry	Earl C.	Flying Instructor	0
Henshall	T.K.	LAC	16
Hesketh	Norman	LAC	03
Hester	N.E.	Flying Instructor	0
Hetherington	L.S.V.	Fl. Lt.; Staff	0
Hibbert	M.J.	LAC	13
Hicks		LAC	03
Hicks	John T.	LAC	17
Higgins		LAC	13
Higgins	T.L.	F/O; Staff	0
Hill, Dr.	Clarence Thomas	Flying Instructor	0
Hill	H. "Harry"	LAC	11
Hill	M.	Flying Instructor	0
Hill	Marvin U.	Link Trainer Instructor	0
Hill	P.E.	LAC	14
Hill	V.	LAC	16
Hill	V.J.	LAC	14
Hill	V.J.	Flying Instructor	0
Himes	L.T. "Lawrence"	AV/C; USAAF	13
Hirst	Derek Albert	LAC	14/15
Hoad	Robert H. "Bob"	LAC	15
Hobbs	Peter	LAC	06
Hobden	A.S.	LAC	17
Hodder	H.B.	LAC	12
Hodgson	A.	LAC	10
Hogan	Mary E.	S.A.A.D.	0
Holgate	J.	LAC	16
Holgate	Maurice M.	LAC	10
Holland		LAC	11
Holley	Aubrey F.	Air Maintenance	0
Holley	G.A.	LAC	12
Holmes		LAC	12
Holmes	Dennis	LAC	11
Holmes	H.C. "Ginger"	LAC	02
Holmes	N.	LAC	07
Hood	Bob	LAC	0
Hooker	J.	LAC	16
Hooker	Eric H.	LAC	14
Hookings	Eric W.	LAC	11
Horlick	Gordon	Flying Instructor	0
Horton	M.G.	LAC	13
Horwood	W.R.	LAC	15
Hosking(s)	R.O.	LAC	11
Hotchkiss	"Les"	LAC	05
Houghton	Spencer	Flying Instructor	0
Houk	Charles Noel	Flying Instructor	0
Houkom	R.	AV/C; USAAF	11
Housley	Christopher "Moose"	LAC	13
Houston	Allan H.	Sgt.	16
Houston	Allan J.	LAC	04
Howard	H.M. Hugh	LAC	13
Howe	J. "Jack"	LAC	08
Howe	Victor D.	LAC	03
Howell	Benton	Office Staff	0
Howerton	William	Air Maintenance	0
Howes	"Bob"	LAC	09
Howlett	D. "Johnny"	LAC	03
Howley	R.J.J.	LAC	03
Hoyle	P.C.	LAC	15
Hoyle	James	Air Maintenance	0
Hoyle	Wallace	Air Maintenance	0
Huckerby	F.W. "Eric"	LAC	08
Huddle	Andrew George	AV/C; USAAF	15
Hudlow	Henry M.	Air Maintenance	0
Hughes	David	LAC	04
Hughes	James H. "Jim"	Flying Instructor	0
Humes	Donald C.	AV/C; USAAF	14
Humphrey		Lt.; USAAF	0
Hunt	J.P.	LAC	08
Hunt	Mary	Mess Hall	0
Huntley	"Gordon"	LAC	16
Huntley	W.G.	LAC	14
Hurley	"John"	LAC	07
Hurst	Lionel	LAC	14/15
Hutchinson	"Happy"	LAC	03
Huyton	H.G.	LAC	13
Hyatt	"Vincent"	LAC	06
Hyde	Eric A.	LAC	13
Hyma[am]s	J.	LAC	08
Hynds	H. Anthony "Tony"	LAC	17
Hyslop		LAC	16
Hytch	W.L. "Dave"	LAC	12
Ibbotson		LAC	05
Igglesden	Mike S.	LAC	10
Ingram	Thomas	LAC	15
Inman	Murray	AV/C; USAAF	16
Inskip	Derek "Dick"	LAC	01
Ireland	"Jimmy"	LAC	07
Irwin	Joe A.	LAC	03
Isles		LAC	15
Izzard	J.	Flying Instructor	0
Jack		LAC	16
Jack	D.	LAC	11
Jackson	Roy Jack	AV/C; USAAF	14
Jacobs	P.	LAC	11
Jacobs	Reginald	Office Staff	0
Jaques	Roland W.	LAC	01
Jarman	Geoffrey A.W.	Fl. Lt.	13
Jarman	George H.	LAC	11
Jarvis	A.N.H.	LAC	12
Jarvis	W.D. "Wally"	LAC	09
Jeans	K.W. "Keith"	LAC	13
Jeffers	T.	LAC	16
Jeffrey	"Widge"	LAC	12
Jeffries		LAC	12
Jeffries	J.L. "John"	AV/C; USAAF	08
Jenkins	J.G.	F/O; Staff	0
Jenkins	J.G. "Johnny" H.R.H	LAC	06
Jenkins	P.E.	LAC	12
Jenner	W.	LAC	17
Jerger	Henry	Flying Instructor	0
Jestes	Robert V.	Flight Dispatcher	0
Johnson		LAC	12
Johnson		AV/C; USAAF	12
Johnson	A.	LAC	10
Johnson	G.J.	LAC	11
Johnson	J.	LAC	11
Johnson	J.B.	LAC	09
Johnson	K.	LAC	03
Johnson	Leslie M.	Air Maintenance	0
Johnson	Robert H.	AV/C; USAAF	13
Johnson	W.E.	Watchman	0
Johnson	William H.	Watchman	0
Johnston	A.G.	LAC	14
Johnston	Charles E.	Air Maintenance	0
Johnston	Cloise E.	Air Maintenance	0
Johnston	J.O.	LAC	08
Johnston	John E.	Janitor	0
Johnstone	"Johnny"	LAC	12
Joliffe		LAC	06
Jones	G.S.	LAC	16
Jones	P.H.I.	Fl. Lt.; Staff	0
Jones	S.C.	LAC	16
Jones	Vincent L.	Grd. School Instructor	0
Jordan	Orley B.	Air Maintenance	0
Joseph	Fred J.	Flying Instructor	0
Joyce	L.	LAC	12
Juckes		LAC	13
Judge		LAC	11
Julier	"Syd"	LAC	07
K	G.H.	LAC	11
Kannewurf	Walter	Flying Instructor	0
Kanter	Emil "Eddie"	LAC	01
Kapell	Luther W.	Flight Dispatcher	0
Kay	David J.	LAC	03
Keane	L.A.	LAC	11
Keech	"Seegoosk"	LAC	04
Keith-Thomas	Anthony "Katie""	LAC	10
Kelly	Thomas U.	Unknown Position	0
Kelson	William Benjamin	Flying Instructor	0
Kemp	R. "Joe"	LAC	12
Kemp	Reginalde Thomas	LAC	15
Kendall	R. "Robert"	LAC	13
Kennedy		AV/C; USAAF	17
Kennedy	J.C. "Craig"	AV/C; USAAF	12
Kenworthy	J.C.	Flying Instructor	0
Kerr	W.J.O.	LAC	14
Kershaw	G.E.	LAC	16
Killicoat	I.M.	LAC	16
King	"Bill"	LAC	04
King	C.R. "Cliff"	LAC	10
King	David C.	AV/C; USAAF	13
King	G.E.	LAC	14
King	Graham	LAC	08
King	R. Douglas W.	LAC	02
Kingdon		LAC	16
Kirk	"Hugh"	LAC	13
Kirrage	G.C.	LAC	12
Kjaer	L.	LAC	12
Klimanek	R.S.P.	LAC	12
Knight	G.	LAC	11
Knott	A.H.	LAC	17
Knott	"Gene"	AV/C; USAAF	05
Knox	Harry A.	Air Maintenance	0
Koerner	Arthur	Air Maintenance	0
Kometer	John F.	Flying Instructor	0
Kominsky	Paul Peter	AV/C; USAAF	15
Kopecky	Joseph	Air Maintenance	0
Kuebler	Don A.	AV/C; USAAF	17
Kugenfeldt		AV/C; USAAF	11
L	A.F.	LAC	11
L "not Laing!"	H.A.	LAC	03
Lacerte		AV/C; USAAF	17
Lacey	Lawrence W.	Flying Instructor	0
Laflin	D.C.	LAC	11
Laird	Helen R.	Office Staff	0
Lambert	James O.	Head Cook	0
Laming	J.	LAC	12
Landless	Jeffrey	LAC	16
Lane	Raymond "Lovers"	LAC	01
Larkin	Harry J.	AV/C; USAAF	14
Larking	Peter C.G.	LAC	13
Larkins	H.W.	LAC	02
Larson		AV/C; USAAF	11
Larson	Frank H.	AV/C; USAAF	13
Laughton	E.R.	LAC	12
Laverock	D. "Dave"	LAC	10
Law	Donald Forbes	LAC	13
Lawking		LAC	12
Lawless	A.E. "Ginger"	LAC	11
Lawrence	Hugh	AV/C; USAAF	17
Lawrence	Jack	LAC	12
Laws	G. "George"	LAC	10
Leach	Clifford Freeman	LAC	15
Leah		LAC	03
Learmouth	Robert Smillie	LAC	15
Ledbetter	James Eudell	Air Maintenance	0
Lee		LAC	17
Lee	"Bob"	LAC	07
Lee	George	LAC	05
Lee	K.B.	LAC	14
Lee	L.A.	Fl. Lt.; Staff	0
Lee	Robert F.	LAC	15
Lees	P.M. "Joe"	LAC	10
Lees	R.A.G.	LAC	14
Lee-Warner	T.J.	LAC	05
LeFeuvre	P.E.	LAC	09
Legard	P. "Pete"	LAC	10
Legate	M.B.	LAC	16
Legg	Norman W. Thomas	LAC	14
Legg	Wayne B.	Air Maintenance	0
Leggett	J.W.	LAC	10

LAST NAME	FIRST NAME	RANK	COURSE NO.
Leiper	A. M. "Jack"	F/O	12
Leonard	John F.	Flying Instructor	0
Lessert	Robert A.	Watchman	0
Lett	J.F.	LAC	10
Leutscher	R.N. "Ron"	LAC	01/2/3/4
LeVieux	Richard S.	AV/C; USAAF	16
Levy	"Frank"	LAC	09
Lewis	Dorr	Air Maintenance	0
Lewis	G.D. "Geoff"	LAC	03/04
Lewis	"Henry"	LAC	08
Lewis	J.A.T.	Office Staff	0
Lewis	J.A.T. "Joe"	Sgt.	14
Lichlyter	W.L. "Jimmy"	AV/C; USAAF	06
Lickorish	Leonard J	LAC	04
Lindberg	George A.	Flying Instructor	0
Lindley	K. Robert H. "Bob"	LAC	13
Lines	Gordon T.	LAC	14
Linquist	Robert E.	Flying Instructor	0
Lipscombe	J.E.	LAC	16
Little	D. "Dave"	LAC	06
Little	T.	LAC	12
Lloyd	"Fred"	LAC	02/03
Lloyd	"Pluto"	LAC	07
London	R.	LAC	11
Londors	A.W.J. "Doug"	LAC	02
Long	Alfred	LAC	02
Long	"Ken"	LAC	04
Longdon	J. "Johnny" "Jimmy"	LAC	06
Looker	John	LAC	16
Loose		LAC	13
Lough	"Pete"	LAC	06
Loughmiller	T.J.	Air Maintenance	0
Love	W.A. "Bill"	LAC	03
Lowson	Leo E.	LAC	01
Lucas	Robert L.	Flying Instructor	0
Ludlow	Dan J.	P/O	16
Luker	Elmer W.	Flying Instructor	0
Lummis	A.	LAC	11
Lupton	J.G.	Fl. Lt.; Staff	0
Lyles	Cecil R.	AV/C; USAAF	14
Lyon	"Mahatma"	LAC	07
M	E.[R] R.	LAC	11
M	J.	LAC	04
M	T.W.	LAC	04
MacDonald	W.	LAC	11
MacDowall	W. "Jock"	LAC	09
Macenis		LAC	12
MacFadzean	R.H.M. "Bob"	LAC	06
Mackay	W.	LAC	16
Mackay	E.D.	LAC	15
Maclennan	Norman "Mac"	LAC	01
Macreadie	D.	LAC	11
Madeley		LAC	16
Magner	W.O.	Janitor	0
Maiden	G.	LAC	08
Maness	M.H.	Air Maintenance	0
Mango	A.S. "Tony"	LAC	02
Manning	J.F. "James" or "Jim"	Wing Com.	12
Mansell-Pleydell	D.G. "Pencil"	LAC	04
Mant	T.	Fl. Sgt. Staff	0
Margerum	C.A.	LAC	02
Maritz	C.S.	LAC	10
Markie	L.W.	LAC	14
Marks	G.A. or J.A.	LAC	16
Marshall	A. "Tony"	LAC	04
Marshall	F.	LAC	05
Marshall	Ray H.	LAC	06
Martin	D.C. "Donald"	AV/C; USAAF	13
Martin	Gene	Flying Instructor	0
Martz	Kathleen	Office Staff	0
Mason	Isaac Glenn	AV/C; USAAF	15
Matthews	D.E.	LAC	09
Matthew	Keith M.	LAC	15
Matthews	R. "Dennis"	LAC	11
Matthews	"Tiger"	AV/C; USAAF	13
Maughan	J.C.	Fl. Lt.; Staff	0
May	A.	LAC	11
McAinsh	"Peter"	LAC	13
McBeath	J.A.	LAC	15
McBee	T.G.	Flying Instructor	0
McBride	E.A.	LAC	14
McCabe	W.	LAC	14
McCall	J.H. "Jock"	LAC	07
McCammon	Chester L.	Air Maintenance	0
McCarley	J.W.	LAC	14
McClain	L.J.	Air Maintenance	0
McCleery	K.D.	AV/C; USAAF	17
McCleod	Roderick	LAC	04
McComb	Robert P. "T-Bone"	Flying Instructor	0
McComiskey	L.	LAC	12
McCord	Robert Morris	AV/C; USAAF	15
McCormack	L.M.	Flying Instructor	0
McCoy	Frank W.	Air Maintenance	0
McCrary	Hugh	AV/C; USAAF	17
McCroskey	Klein L.	Air Maintenance	0
McDermott	Edward P.	Flight Dispatcher	0
McDonald	D.H.	Lt.; USAAF	0
McDonald	Hilton	Air Maintenance	0
McDonald	Thomas K.	Radio Department	0
McDougal	William	LAC	11
McDougal	M.R.	Air Maintenance	0
McFadyen	"Bob"	LAC	12
McFarland	Hiram E.	Air Maintenance	0
McFarlin	R.	LAC	16
McGarry	H. "Mike"	LAC	08
McIntosh	P.C.	LAC	10
McKenzie	J. "Corp"	LAC	07
McKissack		AV/C; USAAF	15
McLaughlin	C.R.	Squadron Leader; Staff	0
McLean	D.	LAC	14
McLennan	T.	LAC	15
McMath	J.E.	LAC	17
McMeekan	W.	LAC	12
McMillan	"Jock"	LAC	06
McNeill	Leonard	LAC	03
McNiece	Leonard L.	Watchman	0
McPhearson	R.J.	Air Maintenance	0
McTaggart	Alex	LAC	01
McVay	C.M.	Flying Instructor	0
McVea	Carroll D.	Canteen	0
Meacock	J.R.	LAC	14
Meade	"Ron" W.	LAC	07
Meek	W. Stewart B.	LAC	13
Melichar	Ernest	Air Maintenance	0
Melling/Mayling	Andrew N. "Andy"	LAC	01
Mellor		LAC	16
Melnicoff	M.	AV/C; USAAF	16
Melville	D.	LAC	08
Mendenhall	E.T.	Air Maintenance	0
Mercer	George S.	LAC	02
Mercy	W. "Bill"	LAC	01/03
Merris	John C.	Flying Instructor	0
Meson	J.J.	LAC	10
Meyers	A.H.	AV/C; USAAF	11
Meyers	Leonard P.	Flying Instructor	0
Meyers	Milton H.	Flying Instructor	0
Michie	Marvin	AV/C; USAAF	13
Miles	P.W. Claude	LAC	12
Miles	James W.	Flying Instructor	0
Millar	I.C.	LAC	13
Miller		LAC	04
Miller	J.G. "Johnny"	LAC	13
Miller	L.	AV/C; USAAF	12
Miller	"Stan"	LAC	12
Millward	C.C.	LAC	14
Millward	J. "Jack"	LAC	06
Milne	J.	LAC	05
Milner		AV/C; USAAF	12
Mingus	Murray Virgil	Flying Instructor	0
Miron	Eddy J.	Flying Instructor	0
Mitchell	"Alf"	LAC	04/05
Mitchell	David J.	Flying Instructor	0
Mitchell	David J.	LAC	11
Mitchell	John J.	Air Maintenance	0
Mitchell	Kenneth H.	LAC	14
Monk	A.J.P.	LAC	13
Monk	B.E.	LAC	16
Montague	J.W.	LAC	14
Montgomery	V.	AV/C; USAAF	12
Moody	E.E.	Air Maintenance	0
Moody	Gilbert T.	Air Maintenance	0
Moore	"Bill"	LAC	05
Moore	M.E.L.	AV/C; USAAF	11
Morgan	Aaron	LAC	03
Morgan	E.J.H. "Jim"	LAC	09
Morgan	Francis Wiley	Flying Instructor	0
Morgan	Frank H.	Ground School Instructor	0
Morgan	G.T.	Air Maintenance	0
Morgan	I.M.P. "Ivor"	LAC	02
Morgan	"Taffy"	LAC	01
Morgan	W.E.H.	LAC	14
Morley	D.	LAC	10
Morley	"Wilf" F.	LAC	03
Morrill	"Barney"	LAC	16
Morris	A.	LAC	17
Morris	Dave	LAC	03
Morris	E.A.G. "Jeep"	LAC	08
Morris	E.R.	LAC	14
Morrish		LAC	17
Morrison	Charles	LAC	05
Morrison	Howard E.	LAC	12
Morrison	J.	LAC	16
Morrison	"Jock"	LAC	04
Morriss	R.E.	Air Maintenance	0
Morse	Robert G.	S.A.A.D.	0
Morton		AV/C; USAAF	16
Morton Smith	M. "Mike"	LAC	09
Mossman	P. Windham M.	LAC	07
Mounger	Joe H.	Flying Instructor	0
Mounsey	H.G.P. "Pete"	LAC	06
Moyle	Joe. E.	LAC	16
Moynier	E.L.	AV/C; USAAF	11
Munger	Joe	Flying Instructor	0
Munt	B.	LAC	11
Murray		Flying Instructor	0
Murray	Jack Wyman	Unknown Position	0
Murton	Phillip G.	Squadron Leader	06
Myers	John C.	Flying Instructor	0
Nabbs	Stanley William	Sgt.	14/15
Nann	R.A.	AV/C; USAAF	17
Nash	Lowanda M.	Canteen	0
Neild	J.A. "Joe"	LAC	09
Neilson	W.A. "William"	LAC	11
Nelson	A.	AV/C; USAAF	16
Nelson	Bruce H.	Air Maintenance	0
Nelson	Mainwaring "Bill"	LAC	02
Nelson	Myrtle H.	Mess Hall	0
Netherton	J.D.	Air Maintenance	0
Nettleton	Frank E.	LAC	03
Neumann	Alfred A.	Flying Instructor	0
Newbold	Reg. V.	LAC	14
Newbury	N.W.D.	LAC	11
Newcomb	"Dick"	LAC	05
Newman	Edward "Ted"	LAC	13
Newsome	Frank	LAC	15
Nicholas	Richard I.A.	LAC	08
Nicholson	A.	LAC	08
Nickolay	"Nick"	AV/C; USAAF	11
Noble	J.B. "John"	LAC	05
Norman	Malcolm C.	LAC	17
Norris	J.J.	LAC	17
Norton	J.A.	LAC	14
Nugent	Johnnie	LAC	04
Nunn	W.H.	AV/C; USAAF	11
Nystrom	Rudolph, Jr.	Flight Dispatcher	0

LAST NAME	FIRST NAME	RANK	COURSE NO.
O	V.E.	LAC	06
Oakley	J.W.	LAC	09
Oates	J. Adrian	P/O	16
O"Callaghan	M.	Fl. Lt.	0
Oden	Elvis R.	Air Maintenance	0
O"Halloran	T	LAC	11
O"Kelley	Jean W.	Air Maintenance	0
Oliver	P.	AV/C; USAAF	12
Oliver	William "Bill"	Flying Instructor	0
Olorenshaw	David H.	LAC	09
O"Neill	A.D.	LAC	14
O"Neill		AV/C; USAAF	13
Orchard		LAC	16
Orr	J.G. "George"	LAC	09/11
Osborne	W.E.P.	LAC	17
Osgerby	D.	LAC	16
O"Shea	B? W?	LAC	12
O"Shea	J.V.	AV/C; USAAF	11
Ottaway		LAC	15
Ouwehand	D.G.	LAC	16
Overhead	W.	LAC	15
Owens	Mel	AV/C; USAAF	16
Page	John Brangwyn	Sgt.	15
Page	H.W.	LAC	17
Page	L.	LAC	12
Page	Seymour "Curley"	LAC	02
Page	Betty	Flight Dispatcher	0
Palk		LAC	15
Palmer	A. "Pat"	LAC	09
Papousek	K. "Carol"	LAC	04
Paradice	C.W. "Charlie"	Sgt.	14
Pardoe	F.A. "Frank"	LAC	11
Park	"Art"	LAC	06
Park	Arthur	Link Trainer Instructor	0
Parker	E.C.	Air Maintenance	0
Parker	Harry Allen	LAC	15
Parker	J.W.G. "Jim"	LAC	03
Parker	Royce Clifton	LAC	15
Parkin	J. "Johnny"	LAC	07
Parnell	Don	LAC	17
Parrish		LAC	17
Parry	Owen A.	LAC	07
Parsonson	Geoff	LAC	17
Partridge	Leslie "Les"	LAC	02
Pate	Robert M.	LAC	16
Patterson	Elsa M.	Mess Hall	0
Pattison		LAC	04
Pattison	"David"	LAC	06
Pattison	David "Herkie"	LAC	07
Patton	James O.	Unknown Position	0
Paul	R.	LAC	11
Paulson	Dennis D.	Air Maintenance	0
Pead		LAC	15
Pearce	David H.	LAC	05
Pearson	J.C.	LAC	10
Penfold	L.A.	LAC	10
Penn	David	LAC	03
Pennycook	A.J.	LAC	17
Pentecost	William "Bill"	Sgt.	16
Perkin	W. "Bill"	LAC	05
Perot	Remy M.	Radio Department	0
Perry	J.H.	AV/C; USAAF	16
Perry	John M.	LAC	01
Peters	Joseph C.	Mess Hall	0
Peterson	L.V.	Air Maintenance	0
Petrie	Charles J.	AV/C; USAAF	14
Philips		LAC	04
Phillips	A.T.	Lt.; USAAF	0
Phillips	G.H.	Squadron Leader; Staff	0
Phillips	Owen W.	LAC	14
Phillips	R.J.C. "Ray"	Corp.; Staff	0
Phillips	Ray M. "Red"	Flying Instructor	0
Phillips	Robert H.	AV/C; USAAF	13
Phillips	S.H. "Sam"	Flying Instructor	0
Phillips	"Tubby"	LAC	09
Phillips	"Whitey"	LAC	07
Philpot	D.H.B.	LAC	14
Pickup		LAC	13
Picton	E.G. "Ted/Blinder"	LAC	01
Pierce	Ben T.	Mess Hall	0
Pierpoint	H.J.	LAC	16
Pile	Len G.	LAC	12
Pinks	Donald S.	LAC	02
Pitcher	Robert "Bob" H.	P/O	15
Pitt	Arthur "Archie"	LAC	07
Plater	E.S.G. "Eddy"	LAC	12
Plimmer	Frank S.	LAC	06
Plotzkar	M.S. "Monty"	LAC	06
Plutt	Opal	Canteen	0
Pope	J.A.R. "Alan"	LAC	02/03
Porter	Wynn W.	AV/C; USAAF	14
Porterfield	Walter H.	AV/C; USAAF	14
Portway	Jack	LAC	04
Povey	J. Les	LAC	05
Powell	H.G.	LAC	10
Powers	Lyle W.	Air Maintenance	0
Power	"Paddy"	LAC	04/05
Prado	Ralph	Air Maintenance	0
Pratt	Alan David	Sgt.	15
Preslensnick	R.L.	Air Maintenance	0
Preston	T.L.	Air Maintenance	0
Price		LAC	03
Price	A.	LAC	11
Price	John Cal.	LAC	17
Price	Johnnie L.	Air Maintenance	0
Price	"Ken"	LAC	06
Price	R.M.	Air Maintenance	0
Price	"Stan"	LAC	07
Priddis	Walter William	AV/C; USAAF	15
Priest	H.J.	Watchman	0
Priestley	F.R.	LAC	06
Prior	"Reg"	LAC	11
Pryke	Stan F.	LAC	14
Puddy	G.S.	LAC	14
Pugh	"Eddy"	LAC	16
Puttick	Gordon W.	LAC	03
Pye	E.C.	LAC	06
Quigley	P. John "Quiggles"	LAC	02
Quilling	T.M.	Air Maintenance	0
Radford	Vernon	LAC	17
Raine	"Ray"	LAC	07
Rake	Derek S.V.	LAC	04
Ralph		LAC	17
Ramshaw	Walter	LAC	01
Ranalow	Patrick B.O." Pat"	LAC	02
Rathbone	J.B.	LAC	14
Rathbone	J.D. "Jonny"	LAC	05
Ratliff	Euel I.	Flying Instructor	0
Ravitts	Richard	AV/C; USAAF	11
Ray	A.	Flying Instructor	0
Ray	Henderson J.	Air Maintenance	0
Reading	C.A.	LAC	14
Redd	Roger A.	LAC	09
Rees	J.W.	Capt.; USAAF	0
Reid	George	LAC	06
Reid	"Johnny"	LAC	07
Reid	Noel	LAC	04
Reynolds		LAC	13
Reynolds		AV/C; USAAF	16
Reynolds	K.	LAC	12
Richards	Pete J.	LAC	04
Richards	Robert C.	LAC	14/15
Rideout	P.W.T.	LAC	14
Ridge	J.	LAC	11
Ridgway	W.L.	LAC	13
Ridler	W. James F. "Jimmy"	LAC	15
Riekie	David M	LAC	11
Ritter	"Tex"	AV/C; USAAF	11
Robbins	M. "Mike"	LAC	06
Roberts	B.R.	Wing Commander	0
Roberts	L.E.	AV/C; USAAF	12
Robins	H.A.	Air Maintenance	0
Robinson	George Clyde	AV/C; USAAF	15
Robinson	J.L.	Flying Instructor	0
Robson	J.	LAC	03
Rodgers	S.A.	LAC	03
Rodriquez	G.G.	Air Maintenance	0
Rogers		LAC	13
Rogers	A.V.	Wing Commander	0
Rolton	Les	LAC	01/02
Rommel	Paul A.	AV/C; USAAF	12
Roork	Vernon C.	Air Maintenance	0
Rose	B.A. "Lefty"	LAC	13
Rose	Osceola	Air Maintenance	0
Ross	J.F.	AV/C; USAAF	12
Ross	J.H. "Johnny"	LAC	08
Rose	M.H.	LAC	11
Ross	Mike	LAC	04
Rowlands	"Tiny"	LAC	06
Rowley	R.A.	LAC	17
Rowley	Sid	LAC	16
Roy	Shelton H.	Flying Instructor	0
Roy	W.	LAC	11
Rubenfeldt	M.	AV/C; USAAF	11
Rugen	H.	LAC	16
Runnels	Art. E., Jr.	Flight Dispatcher	0
Runnels	Wilburt	Link Trainer Instructor	0
Russ	O.P.	Flying Instructor	0
Russell		LAC	09
Russell	P.G.	Air Maintenance	0
Rutz	Melvin F.	Air Maintenance	0
Ryall		LAC	13
S.	J.W.G.	LAC	07
Saegert	A.E.	Lt.; USAAF	0
Saidler	J.D. "Jimmy"	LAC	07
Salkin	Alf	LAC	14
Salyers	Otha H.	Flying Instructor	0
Samson	G.N.	AV/C; USAAF	17
Sandberg	Bruno W.	Unknown Position	0
Sanders	G.E. "Geoff"	LAC	10
Sanders	L.J. "Leo"	AV/C; USAAF	13
Sanderson	Frederick	LAC	03
Sandland	C.B.	LAC	10
Saul	"Tom"	LAC	07
Sawyer	H.D. "Tom"	LAC	01
Sayer	Gordon "Grandpa"	LAC	01
Schellenberger	Ralph	Flying Instructor	0
Schmitt	Hugh L.	AV/C; USAAF	14
Schofield		LAC	16
Schuchman	V.H.	Air Maintenance	0
Schultz	Carl R.	AV/C; USAAF	14
Scillitoe	D.L.	LAC	14
Scott	E.W.	AV/C; USAAF	12
Scott	H.G. "Hugh"	LAC	05
Scott	Robert	LAC	06
Scovill	L.D.	Air Maintenance	0
Seabrook		LAC	13
Sealey	David K.	LAC	14
Seamans	Robert T.	Flying Instructor	0
Seaman	Vernon C.	Air Maintenance	0
Searle	H.D. "Ginger"	LAC	02
Searles	E.	LAC	11
Searles	V.J.	LAC	17
Seddon	H.W.A.	LAC	08
Seeley	J.T.	LAC	10
Segelquist	J.	AV/C; USAAF	11
Sellars	Geoff	LAC	17
Settlemeyer	Claude H.	AV/C; USAAF	07
Shadbolt		LAC	17
Shamback	L.A.	LAC	11
Sharp	Edgar V.	Janitor	0
Sharpe	J.S.B.	LAC	15
Sharpe	"Tom"	LAC	07
Shattock	David H.	LAC	15
Shea	George Raymond	Fl. Lt.; Staff	0
Shelley	A. "Gus"	LAC	01
Sherman		AV/C; USAAF	12
Sherwood	Anthony J.	LAC	17

LAST NAME	FIRST NAME	RANK	COURSE NO.
Sherwood	Anthony "Tony" J.	LAC	16
Sherwood	W.V.	LAC	17
Shipman	J.J., Lt.	S.A.A.D.	0
Shopneck	Sam Milton	AV/C; USAAF	15
Short	A.J.	LAC	13
Showalter	Lehn H.	Flying Instructor	0
Showman	D.	AV/C; USAAF	12
Shultz	Clark	Air Maintenance	0
Siegfried	B. "Bern"	LAC	16
Sillitoe	K.	LAC	01
Sim		LAC	17
Simmins	H.D.	LAC	07
Simmonds		LAC	03
Simmonds	H.E.	LAC	17
Simmonds	S.G.	LAC	17
Simmons	H.D.	LAC	14
Simmons	L.E.	Flying Instructor	0
Simmons	Russell Ben	Unknown Position	0
Simpson	F.W.H.	Fl. Lt.; Staff	0
Simpson	Fred	LAC	04
Simpson	K.W.	Group Captain	11
Simpson	Kenneth W.	LAC	13
Simpson	Stewart B.	LAC	13
Sims	Albert	LAC	07
Sims	L.	Flying Instructor	0
Singer		LAC	11
Singleton	J.	LAC	11
Siveter		LAC	08
Skerman	"Harry"	LAC	17
Slade	A.E.	LAC	11
Slavin	F.J.	Air Maintenance	0
Slavin	George R.	AV/C; USAAF	14
Smart	K. Neville	LAC	03
Smiley	Earl H.	Air Maintenance	0
Smit	J.C.H.	LAC	13/14
Smith	A. "Alan"	LAC	11
Smith	C.H.	Air Maintenance	0
Smith	Fred H.	Air Maintenance	0
Smith	Isaac Alfred	Unknown Position	13
Smith	"John"	LAC	07
Smith	John T.	Flying Instructor	0
Smith	Leighton D.	Flying Instructor	0
Smith	M.A.K. "Kenneth" A.	LAC	13
Smith	Micael Morton	LAC	09
Smith	not K.	LAC	13
Smith	Olive	Mess Hall	0
Smith	P.L.	Air Maintenance	0
Smith	Percy	LAC	03
Smith	Peter W.	LAC	14/15
Smith	S.M.	LAC	10
Smyth-Piggott	Hugh	LAC	04/05
Snell	Anthony "Tony"	LAC	01
Snow	Forde C.	Watchman	0
Snowden	"Roy"	LAC	17
Softley	F.P.	LAC	17
Soloman		LAC	04
Sommovigo	Marcello J.	AV/C; USAAF	11
Sorrells	C.H.	Air Maintenance	0
Soule	F.E.	Air Maintenance	0
Southfield		LAC	11
Sowerbutts	John	LAC	03
Spain	David S.	Flying Instructor	0
Spencer	A.M. E. "Bert"	LAC	16
Spencer	"Pete" Peter	LAC	03/04
Sponsler	Russ E.	Air Maintenance	0
Spooner	H.	LAC	10
Spooner	"Jock"	LAC	07
Spore	Eletha	Office Staff	0
Sprent	W.S. "Bill"	LAC	01
Stabler	Eric	LAC	12
Staley	F.	LAC	14
Stanfield	H.W. "Harry"	LAC	09
Stanley	E.D.	LAC	14
Stanley	Howard	Parachute Department	0
Stanning	M.R.	LAC	09
Stark	R.W.	AV/C; USAAF	15
Stedman	L.W.	LAC	16
Steedman	Alex	LAC	11
Steel	J. Harvey	LAC	07
Steele	Donald W.	Ground School Instructor	0
Steele	J.R. "Jock"	LAC	09
Stephens	Eric	LAC	04
Stephens	F.S.	LAC	14
Stephens	J.F. "John"	LAC	13
Stephens	K.J.	LAC	09
Sterling	Raymond	Air Maintenance	0
Stevens		LAC	13
Stevens	Eric	LAC	03
Stevens	Fred R.	LAC	17
Stevens	T.J.A.	LAC	16
Stewart	Andy	LAC	01/02
Stewart	Ray H.	Air Maintenance	0
Stieglmar	Marion L.	Flying Instructor	0
Stirling	R.W.	AV/C; USAAF	16
Stone	Dennis	LAC	17
Stoneley	Albert Clifford	Sgt.; LAC	15
Stowe	James R.	Watchman	0
Stratford	D.W. "Don"	LAC	07
Stratton	J.W.	LAC	10
Strong		LAC	16
Stubbs	"Stoker"	LAC	07
Sturgeon	William M.	Unknown Position	0
Sturges	R.M.	Air Maintenance	0
Such	George	LAC	05
Sullivan	M.W. or N.W.	LAC	16
Sullivan	W.B.	AV/C; USAAF	13
Sunderland	H.	LAC	14
Suren	H.F.	Squadron Leader; Staff	0
Suriano	F.A.	Air Maintenance	0
Survage	E.	LAC	12
Sutherland	C.	LAC	10
Sutherland	J.W. "Jack"	LAC	02
Sutherst	"Jack"	LAC	05
Swain	John F.	LAC	07
Swallen	Robert E.	AV/C; USAAF	16
Swan		LAC	17
Swearingen	A.O.	Air Maintenance	0
Swenson	Norman W.	Flying Instructor	0
Swinehart	William Keith	AV/C; USAAF	15
Swinnerton	Don G.	LAC	13
Symons	"Tom" or "Taffy""	LAC	04
Tait	A.J.	LAC	09
Talbot	E. William	LAC	01
Tatlock	Jack	LAC	15
Taylor	Amret M.	Flying Instructor	0
Taylor	Archie R.	AV/C; USAAF	12
Taylor	Bryan W.	LAC	16
Taylor	G.	LAC	08
Taylor	George	LAC	01/02
Taylor	Thomas D.	Squadron Leader	03
Taylor	W.	LAC	05
Telvin	K.B.	LAC	14
Temple	Albert C. or D.	LAC	02
Tennant	"Dormouse"	LAC	07
Tester	A.H.	LAC	01
Thatcher	E.G.	LAC	04
Thomas	D.O. "Dan"	LAC	01
Thomas	E.W.P.	LAC	08
Thomas	I.	LAC	08
Thomas	R.A.	LAC	15
Thompson	E.	LAC	15
Thompson	E.O.	LAC	14
Thompson	G. "Geoff"	LAC	13
Thompson	J.	LAC	11
Thompson	J.B.	LAC	05
Thompson	J.E.	LAC	17
Thompson	J.W.	LAC	05
Thompson	Rex B.	Air Maintenance	0
Thompson	"Tommy"	AV/C; USAAF	05
Thorndick		LAC	17
Thorne	James Neale "Jim"	AV/C; USAAF	07
Thorne	K.W.	LAC	15
Thornton	R.	LAC	16
Thorogood	Ray	LAC	05
Tietz	Carl G.	Air Maintenance	0
Timperley	Leonard Joseph	LAC	13
Titchener	Geoff A.	LAC	15
Titcomb		LAC	15
Tomlinson	Douglas F.M.	Sgt.	16
Treen	J.V.	Fl. Sgt.	0
Treigle	L.W.	LAC	14
Trimmer	Frank	Flying Instructor	0
Trinder	I.F.	LAC	14
Turnbull	T.P. "Toggers"	LAC	04
Turnbull	William Longstaff	LAC	15
Turner	C.L.	Fl. Lt.; Staff	0
Turner	E.H.C.	LAC	12
Turner	"Joe" Altham	LAC	12
Turner	Lester	Air Maintenance	0
Turner	Terry	LAC	01
Turpin		LAC	17
Tuttle	Shelby G.	AV/C; USAAF	12
Twigger	A. Ray	LAC	15
Twomey	T.M. "Terry"	LAC	02
Twomey	Vincent H.	LAC	01
Tye		LAC	17
Tyson	"Mike"	AV/C; USAAF	16
Udove	John A.	Grd. School Instructor	0
Underwood	H.G. or N.G.	LAC	13
Urch	Stan Syd	LAC	11
Ustrell	A.R.	LAC	17
Utberg	Neil	Link Trainer Instructor	0
Vaden	Walter L.	Air Maintenance	0
Van Norman	Melvin Forrest	AV/C; USAAF	15
Vaughan	William B.	AV/C; USAAF	14
Vickery	Les "Wings"	LAC	07
Victor	Paul F.	Flying Instructor	0
Vincent	B.J.	LAC	13
Vinnidge	Barton P.	Flight Dispatcher	0
Vowles		LAC	04/05
Vowles	Eric	LAC	12
Vreauls	W.E.	AV/C; USAAF	17
W	D.	LAC	11
Wagstaff(e)	Richard F. "Dick"	LAC	15
Wainwright	J.	LAC	08
Wakeman	Ronald	LAC	08
Walker	J. Barron "Johnnie"	LAC	08
Walker	J.C.	Flight Dispatcher	0
Wallis	Marc H. "R."	LAC	02
Walsh	Guy C.H.	LAC	02
Walters	P.A.	LAC	13
Waltman	William E.	Mess Hall	0
Walton		LAC	13
Walton	L. "Lew"	LAC	09
Walton	Ronald D.	LAC	04
Wampler	V.M.	Flying Instructor	0
Ward	Billy W.	Flying Instructor	0
Ward	J.	LAC	16
Ward	J.C.	Air Maintenance	0
Ward	Jack	LAC	14
Ward	Stan W.	LAC	12/13
Ward	Billy W.	Flying Instructor	0
Warman	W.H. "Bill"	LAC	02
Warner	Jim "Jack" Lee	LAC	01
Warren	A.M.	AV/C; USAAF	17
Waters	Stanley	LAC	16
Watkins	Dr. William V.	AV/C; USAAF	17
Watson	A.N.	LAC	11
Watson	D.G. "David"	LAC	13
Watson	David	LAC	03
Watson	Ferrell E.	Air Maintenance	0
Watson	J.G.	LAC	05
Watson	J.H.G. "Joe"	LAC	04
Watson	Jack S.	LAC	02
Watson-Taylor	Brian	LAC	16/17

LAST NAME	FIRST NAME	RANK	COURSE NO.
Watters	L. or S.A.	LAC	16
Weatherford	J. S.	Flying Instructor	0
Weaver	Everett B.	Unknown Position	0
Weaver	R.	LAC	08
Webster	Brian F.	LAC	13
Webster	L.	LAC	14
Webster	T.H.	LAC	17
Weeks	John	Air Maintenance	0
Weeks	Peter G.	LAC	05
Weems	G.	AV/C; USAAF	12
Weller	K.C.	LAC	05
Welsch	G.W. "Jack"	Flying Instructor	0
West	Carl	Air Maintenance	0
West	John	LAC	03
West	P. Arthur	LAC	02
Westcott	K.	LAC	17
Weston	"Bill"	LAC	17
Weston	David B.S.	LAC	14/15
Whalley	L. Stanley "Stan"	LAC	0
Wharton	George	LAC	01
Wheeler		LAC	16
Wheeler	R.W. "Ron"	LAC	07
Whitaker	Robert	Air Maintenance	0
White	E.J.	LAC	13
White	J.H.	LAC	17
Whitely	"Jack"	LAC	05
Whitmore	George Hayden	Unknown Position	0
Whittaker	Rowland "Ron" C.	LAC	02
Wieder	John	Flying Instructor	0
Wignall	L. "Les"	LAC	05
Wilcox	J.F.O.	LAC	16
Wilda	Wesley	Air Maintenance	0
Wilett	F.T.	Office Staff	0
Wilkins	R.E.	LAC	14
Wilkinson	B.C.	LAC	16
Wilkinson	"Bill"	LAC	04
Willett	D.J.	LAC	13
Willett	F.T.	AV/C; USAAF	11
Williams	D.C.	LAC	17
Williams	Edmund H.	LAC	08
Williams	Graham P.	LAC	10
Williams	Lucretia	Canteen	0
Williams	Mabel D.	Canteen	0
Williams	Merle M.	Flying Instructor	0
Williams	R.A.	LAC	10
Williams	Rovern	Air Maintenance	0
Williams	W.K. "Bill"	LAC	03
Willmott	John "Jerry" N.T.	LAC	17
Wilma	R.	AV/C; USAAF	11
Wilson	E.F.	LAC	03
Wilson	J.G.	LAC	07
Wilson	J.H.	LAC	16
Wilson	S.F.	LAC	14
Wilson	Walker	Air Maintenance	0
Windle	Robert	LAC	01
Wiseman	C.A.	AV/C; USAAF	17
Wolf	Lawrence M.	Flying Instructor	0
Wood		LAC	16
Wood	Charles Franklin	AV/C; USAAF	15
Wood	John	LAC	02
Wood	John "Weaver"	LAC	03
Wood	T.W.K.	LAC	17
Woodfin	S.D.	LAC	16
Woodger	C. "Bob"	LAC	08
Woodmass	Stan	LAC	17
Woods	"Grandma"	LAC	07
Woods	H.	LAC	08
Woodwards	Robert G. "Bob"	Sgt.	14
Woor		LAC	16
Wootton	Clarence J.	Flying Instructor	0
Worrall	O. Gerald	LAC	13
Worthington		LAC	16
Wright		LAC	13
Wright		LAC	16
Wright	G.	LAC	08
Wright	Leslie	LAC	02
Wright	Peter Gilbert	LAC	15
Wroe	"Tommy"	LAC	06
Wylie	Kelsie M.	Air Maintenance	0
Yearrill		LAC	16
Youens		LAC	17
Youens	Jack	LAC	03
Young	Robert Lee	Air Maintenance	0
Young	"Scot"	LAC	07
Yule	D.F.	LAC	14
Zanzottera	J.H. "Joe"	LAC	08
Zerba	Lee R.	AV/C; USAAF	12
Zied	P.	LAC	16

Thhis aerial photo shows the Darr School of Aeronautics. *Courtesy Bret Carter*

— SPARTAN SCHOOL OF AERONAUTICS · MIAMI, OKLAHOMA —

LAST NAME	FIRST NAME	RANK	COURSE NO.
Abel	S.	LAC	15
Acott	Wallace A.	AC2	21
Adam	A.W.	AC2	23
Adams	M.	AC2	23
Adamson	Deynee	Staff	0
Adderley	A.	Squadron Leader; Ground School Sup.	0
Alexander	J.O.		26
Allan	R.B.	AC2	25
Allan	W.J.	LAC	15
Allen	J.G.	LAC	18
Allen	J.O.	LAC	17
Allen	R.	LAC	17
Allen	Spencer G.	LAC	19
Allen	W.A.	LAC	10
Allin	P.S.	LAC	17
Allman	H.R.	LAC	16
Amey	R.L.	LAC	16
Anderegg	Henry Harold	Adv. Flight Instructor	0
Anderson	J.	AC2	22
Anderson	L.O.	AV/C; USAAF	18
Anderson	W.C.	LAC	19
Andrews	C.W.	LAC	19
Andrews	G.R.	LAC	10
Angus	W.C.H.	AC2	23
Appleby	George F.	AC2	22
Appleton	J.A.	LAC	12
Argodale	John	Link Trainer Instructor	0
Artes	E.J.	AC2	24
Arthur	L.	SGT	14
Ash	B.E.	LAC	18
Askew	W.L.	W/O; USAAF	0
Astbury	John K.		27
Atherton	J.S.	LAC	18
Atkinson	H.T.G.	LAC	13
Atkinson	M.T.G.	CPL	12
Austin	G.F.	LAC	16
Axton	R.H.	LAC	11
Badwey	John	Field Dispatcher	15
Bailey	K.A.	LAC	15
Bailey	Peggy	Ground School Staff	0
Bailey	T.	AC2	21
Bain	J.G.	AC2	25
Baker	A.V.J.	AC2	20
Baker	Billie	Staff	0
Baker	D.G.	LAC	19
Baker	M.H.	AV/C; USAAF	19
Baker	Tony		26
Baldwin	Allan D.	LAC	15
Bale	W.J.	LAC	13
Balfour	Maxwell	Spartan CEO	0
Ballinger	R.A.	AC2	22
Bamber	R.R.	AC2	21
Bamford	G.P.	LAC	18
Bampton	A.L.	AC2	21
Bamsey	W.L. Bill	CPL	13
Banahan	J.D.P.	LAC	12
Banks	Anthony F.	LAC	19
Banks	Reginald	LAC	18
Banning	L.J.	AC2	22
Banton	J.	LAC	1
Banyard	C.W.	LAC	15
Barden	John	LAC	20
Barger	Ruth	Staff	0
Barker	Stan	Ground School Instructor	27
Barker	Max S.	Unknown Position	0
Barker	J.B.	LAC	4
Barlow	N.H.	Cpl.	0
Barnett	G.M.	AC2	22
Barnum	Bouford E.	Flight Instructor	0
Barr	D.	LAC	12
Barr	M.B.	LAC	15
Bartlett	A.G.	LAC	12
Bartlett	M.T.O.	LAC	4
Base	Keith M.	CPL	15
Bass	E.G.	LAC	6
Bass	Ralph A.	AV/C; USAAF	15
Bassett	R.M.	LAC	18
Bassett	Stonewall	Adv. Flight Instructor	0
Bassford	Jim	LAC	19
Bate	A.T.	AC2	24
Bate	W.T.	AC2	21
Bateman	A.C.L.	AC2	23
Bateson	P.G.	AC2	24
Bath	A.A.	LAC	15
Battin	D.A.W.	LAC	7
Batty	F.J.	LAC	18
Bauer	Ed	Staff	0
Baxendale	E.	AC2	21
Baxter	C.P.	AV/C; USAAF	18
Baxter	Raymond	LAC	5
Bean	Amos L.	Maintenance	0
Beasant	E.A.	LAC	10
Beatty	Edson W.	Adv. Flight Instructor	0
Beatty	William W.	Staff	0
Beaty	Amer	Bus Driver	0
Beaumont	F.	Flight Instructor	0
Becker	Gordon W.	Flight Instructor; Col/AF Ret.	0
Bedell	Frederick J.	Adv. Flight Commamder	0
Bedford	A.	AC2	21
Beedle	D.H.	LAC	14
Beeler	W.C.	AV/C; USAAF	14
Beesley	P.L.	LAC	2
Beeson	C.T.G.	AC2	23
Bell	Harry W.J.	AC2	21
Bell	J.	AC2	25
Bell	W.	Sgt.	0
Bellenger	A.B.	LAC	19
Belliad	W.S.	LAC	15
Belsen	Henry	Air Maintenance	0
Benson	A.C.	LAC	18
Benson	P.W.	AC2	24
Benton	Cyril	LAC	19
Berg	R.	AC2	24
Bergmann	Albert Francis "Al"	Adv. Flight Instructor	0
Bergmann	Leona Helen	Para Rigger	0
Berndt	Ed	Staff	0
Berry	Clyde D.	Primary Flight Instructor	0
Berry	S.	AC2	20
Berryman	Brian A.C.	LAC	18
Best	A.C.	LAC	6
Beveridge	A.	Squadron Leader	0
Bevis	J.O.	AC2	23
Bickham	W.D.	AV/C; USAAF	17
Biggs	S.C.	AC2	21
Bilsland	J.	AC2	25
Bilsland	J.D.	LAC	13
Binns	G.E.	SGT	24
Birch	Ron F.	LAC	2
Bird	R.H.	LAC	1
Bishop	C.W.	AC2	23
Bissett	J.	AC2	21
Black	R.L.	AC2	25
Blackwell	H.J.	AC2	22
Blain	Robert	Staff	0
Blair	R.A.	AV/C; USAAF	19
Blakey	W.V.	LAC	7
Block	F.	LAC	2
Blogg	G.E.	LAC	18
Blome	Arlind	Tech. Sgt. USSAF Clerk	0
Bloy	H. Bert	AC2	20
Blyton	A.H.	AC2	23
Bollar	Harry E. "Hap"	Adv. Flight Instructor	0
Bolton Smith	F.R.	LAC	6
Bonning	Albert W.	Fl. Lt.	0
Bonny	G.L.	CPL	25
Borrows	K.A.	AC2	22
Borthwick	William	AC2	25
Boucher	E.G.	LAC	2
Boulet	T.	AV/C; USAAF	18
Bounds	P.	AC2	23
Bower	J.B.W.	AC2	20
Bowes	Alan	AC2	25
Bowyer	A.R.	LAC	17
Boyd	D.	LAC	19
Boyd	J.D.	AV/C; USAAF	18
Boyer	Robert Kerr	Flight Instructor	0
Bradbury	G.	LAC	14
Bradfield	Colin		26
Bradford	C.E.	AV/C; USAAF	19
Bradshaw	Jack T.	LAC	9
Brady	R.M.	LAC	23
Bramwells	H.G.	LAC	8
Brand	S.S.	LAC	4
Brant	T.W.	AC2	25
Bray	R.C.	AC2	21
Bredengerd	John J.	Staff	0
Bree	James T.	LAC	14
Brewer	L.	LAC	15
Brewington	Russ D.	AV/C; USAAF	17
Bridge	C.C.	Staff	0
Brierley	John F.	LAC	17
Briggs	D.A.	Squadron Leader	0
Brighton	L.J.	AC2	22
Brink	Marion Jack	Air Maintenance	0
Brisdon	S.C.	AC2	23
Brittain	C.J.	LAC	6
Brittain	D.F.	LAC	12
Broadbent	Geoffrey	LAC	16
Brockbank	C.A.V.	LAC	19
Brockway	Richard	Flight Commander	0
Bromwell	D.G.	LAC	14
Bromwich	H.A.	AC2	24
Brookshire	R. H.	AV/C; USAAF	17
Brown	F.J.	LAC	7
Brown	G.C.	AC2	20
Brown	G.E.	AC2	23
Brown	J.B.	LAC	9
Brown	J.L.	LAC	2
Brown	Nichol	LAC	14
Brown	R.	LAC	3
Brown, Jr.	Romie Gaines	Air Maintenance	0
Brown	W.	LAC	17
Brown	Walter	Staff	0
Browne	W.N.	LAC	12
Browne	B.R.	LAC	13
Browne	K.	LAC	2
Bruce	Danny	LAC	19
Bruce	J.J.	LAC	13
Bruce	P.E.	LAC	3
Bruming	W.C.	LAC	23
Brunskill	E.	LAC	1
Brunswick	Roy E.		26
Bryan	W.R.	AC2	24
Bryant	C.J.	LAC	5
Bryce	W.	AC2	25
Buck	O.C.	LAC	14
Buck	Richard L.	AV/C; USAAF	18
Bucklow	W.	LAC	16
Buglass	J.A.	LAC	19
Bull	G.G.	LAC	19
Bumford	Evan C.	LAC	5
Bumpus	S.W.	AC2	24
Burgess	Burl N.	Air Maintenance	0
Burn	A.E.	AC2	22

LAST NAME	FIRST NAME	RANK	COURSE NO.
Burnell	S.J.	AC2	23
Burns	N.W.G.	LAC	9
Burrow	G.R.F.	LAC	17
Burt	Peter C.	LAC	13
Burtenshaw	P.J.	AC2	23
Burton	R.G.	AC2	20
Burvill	K.I.	AC2	25
Burwood	G.H.S.	LAC	13
Busbridge	A.H.	LAC	13
Bush	E.E.	AV/C; USAAF	14
Bushen	H.	LAC	6
Bustle	Wayne W.	US Army Medical Corp.	0
Butcher	E. Terry	AC2	21
Butler	B.	LAC	18
Butler	C.J.	CPL	8
Butler	E.G.	LAC	2
Butler	Ivor E.S.	LAC	19
Butlin	W.J.	LAC	15
Butlor	M.C.	LAC	19
Butt	Jack	AC2	20
Button	D.A.	AC2	23
Buttress	B.C.	LAC	16
Byrne	J.K. "Paddy"	LAC	11
Cadman	C.	AC2	22
Cady	R.	AC2	21
Cameron	A.H.	LAC	13
Cameron	A.I.	LAC	15
Cameron	D.S. Bud	LAC	17
Campbell	A.	LAC	6
Campbell	D.B.	LAC	15
Campbell	G.M.	LAC	3
Campbell	H.	LAC	6
Campbell	P.W.G.	AC2	24
Cardwell	F.	CPL	13
Care	R.	AC2	23
Carn	Glen	Staff	0
Carn	N.G.G.	LAC	11
Carr	James A.	Flight Instructor	0
Carter	Eric "Nick" H.	LAC	20
Carver	Douglas	LAC	19
Casey	T. Jack	Chief Flying Inst. US	0
Cash	Dennis	LAC	19
Castledine	R.	LAC	16
Cathcart	H.T.	AC2	21
Cepeling	R.R.B.	AC2	23
Challis	D.M.	LAC	14
Chalmers	D.	LAC	6
Chamberlin	R.E.	LAC	7
Chambers	Carl B.	Staff	0
Champion	C.N.	AC2	20
Chancellor	W.C.	AV/C; USAAF	14
Chandler	Max	Field Dispatcher	0
Chapman	F.A.	LAC	14
Chapman	F.J.	AC2	22
Charbonnet	P.	AV/C; USAAF	15
Charlesworth	A.	LAC	19
Charleton	L.F.	AC2	25
Charnock	E.R.	AC2	24
Chase	G.B.S.	LAC	17
Chase	John R., Jr.	AV/C; USAAF	16
Chase	Peter E.	Adv. Flight Instructor	0
Chasemore	R.A.	Squadron Leader	17
Cheatham	Glen	Staff	0
Checkley	R.W.	LAC	15
Chedlow	Barry W.	LAC	6
Cheesman	B.R.	LAC	13
Chilton	R.G.	AC2	23
Chisnall	B.N.	LAC	15
Church	P.A.	AC2	25
Clark	C.A.	LAC	10
Clark	H.E.W.	AC2	24
Clark	Richard C.	AV/C; USAAF	15
Clark	T.H.	US	11
Clark	W.J.A.	AC2	21
Clarke	A.R.	LAC	10
Clarke	H.	AC2	23
Clarke	Harold	LAC	24
Clarke	John C.	LAC	6
Clarke	M.F.	AC2	23
Clarke	R.N.		27
Clarkson	A.P.	Flight Instructor	0
Clawson	Everett L.	Air Maintenance	0
Clay	S.W.A.	LAC	19
Clayton	G.	LAC	14
Cliffe	H.L.	AC2	24
Cobb	Henry Clay	Flight Instructor	0
Cobbett	S.W.	LAC	18
Cochran	Maurice E.	Air Maintenance	0
Cocking	W.J.	Fl. Lt.	0
Cody	C.T.	Wing Commander	1
Cody	Marian	Link Trainer Instructor	0
Colegate	J.	LAC	6
Colegate	John	LAC	5
Coleman	B.R.J.	AC2	23
Coleman	Les A.	AV/C; USAAF	15
Coleshill	G.H.E.	LAC	9
Collier	J.G.	AC2	24
Collinge	H.A.	AC2	25
Collins	W.J.	AC2	24
Colquhoun	D.W.	LAC	15
Coltart	J.	AC2	22
Connolly	F.W.	LAC	14
Connolly	J.W.	LAC	8
Constable	George E.	AV/C; USAAF	18
Cook	A.M.	LAC	6
Cook	A.R.	LAC	3
Cook	Jackson Warren	Flight Dispatcher	0
Cook	W.L.	LAC	19
Cooke	C.K.	Ground Captain	15
Cooke	J.	LAC	8
Cooke	N.A.	AC2	20
Cooke	W.R. "Bill"	LAC	7
Coombes	Norman R.	AC2	24
Cooper	A.A.E.	LAC	15
Cooper	John		26
Cooper	K.V.	AC2	20
Cooper	L.R.	AC2	25
Cooper	W.K.	LAC	17
Cooper	W.K.	AC2	21
Coote	E.W.	LAC	6
Copeland	R.W.	LAC	12
Corbett	Eddie A.	LAC	13
Corbett	J.M.	AC2	20
Cordell	R.L.	AC2	22
Cornell	E.R.	LAC	6
Cornish	R.C.	AC2	21
Cory	W.A.	LAC	9
Cotton	G.W.	AC2	22
Cottrell	W. George E.	LAC	20
Couchman	R.A.	Fl. Lt.	0
Coult	Tony		26
Courtnay	E.R.	LAC	6
Coverdale	Roy E.	LAC	9
Covington	W.T.	LAC	3
Cowdrey	H.J.	LAC	16
Cowie	Jim		27
Cowley	H.	LAC	7
Cowling	F.J.	AC2	25
Cowood	C.	LAC	10
Cox	A.	LAC	14
Cox	F.P.	Squadron Leader	8
Cox	F.W.	LAC	8
Cox	Gordon C.		26
Crabtree	P.J.	LAC	13
Craig	L.W.H.	LAC	15
Craig	W.K.	LAC	13
Crane	John H.	LAC	17
Crate	G. L.	AC2	23
Cressler	Edward P.	Flight Instructor	0
Crisp	N.J.	LAC	20
Croft	P.F.	LAC	16
Crowther	J.E.	LAC	18
Cruse	D.E.	LAC	1
Cullen	M.H.	LAC	11
Culmer	A.W.	AC2	22
Cumming	W.S.	LAC	19
Cunliffe	Geoffrey H.	LAC	12
Cunliffe	L.	AC2	23
Curlee	Eleanor	Staff	0
Curling	D.J.	LAC	1
Curtis	Clarence	AC2	23
Cuthbert	J.	AC2	24
Cutmore	R.F.	LAC	15
Daffendoll	Jess Roy	Adv. Flight Instructor	0
Dahl	Melvin	Air Maintenance	0
Daley	Wayne Eugene	Air Maintenance	0
Dalgliesh	J.H.	AC2	23
Dallen	J.G.	LAC	12
Dalrlymple	A.	LAC	1
Daly	Ronald G.	LAC	6
Dalzell	R.C.	LAC	7
Damon	Henry E.	AV/C; USAAF	19
Daniel	"Slim"	Staff	0
Darr	Ned Robertson	Adv. Flight Instructor	0
DaVee	Billy	Stenographer	0
Davidson	A.	AC2	20
Davidson	Harry Lawrence	Flight Instructor	0
Davidson	J.	SGT	7
Davies	A.I.	LAC	6
Davies	C.H.	LAC	1
Davies	D.D.	LAC	4
Davies	D.W.	LAC	2
Davies	G.	LAC	13
Davis	A.R.N.	LAC	4
Davis	C.W.J.	LAC	17
Davis	E.G.	LAC	15
Davis	Edward G.	Flight Instructor	0
Davis	Frederick Payson	Flight Instructor	0
Davis	G.P. "Dave"	LAC	16
Davis	Harvard W.	Adv.Flight Instructor	0
Davis	K.C.	LAC	13
Davis	J.L.	LAC	16
Davis	Paul	Staff	0
Davis	Peter		26
Davis	R.C.	LAC	14
Dawson	H.	LAC	10
Day	H.H.	LAC	6
Dear	D.S.	Flight Instructor	0
Deas	G.C.T.	LAC	10
Death	R.F.	CPL	10
DeLours		Adv. Flight Instructor	0
Denby	P.	AC2	23
Dennis	Charles G.	AV/C; USAAF	14
Derrett	Brian L.	LAC	8
Deuntzer	D.C.	Fl. Lt.	0
Devlin	P.J.	LAC	12
DeWald	Robert Michael	Flight Instructor	0
DeWatters	W.G.	Flight Instructor	0
Dewdney	F.	LAC	16
Di Felice	V.P.	LAC	19
Dibb	D.A.	AC2	25
Dibbell	Archie T.J.	AC2	21
Dickenson	Ken	LAC	16
Dickenson	W.	LAC	18
Dickey	D.O.	LAC	14
Dickie	W.A.	LAC	16
Dickinson	James E.	AV/C; USAAF	18
Dickinson	K.H.	LAC	16
Diment	W.H.	F/SGT	14
Dinnie	David	LAC	18
DiSalvo	T.	AV/C; USAAF	15
Disbury	G.F.	LAC	4
Dixey	E. Arthur	LAC	19
Dobson	A.	LAC	9
Dobson	Michael F.H.	Squadron Leader	19

LAST NAME	FIRST NAME	RANK	COURSE NO.
Doggrell	W.J.	AC2	23
Doig	D.T.	LAC	15
Dolling	G.F.	AC2	22
Dolphin	Harry A.	LAC	19
Donahue	W.R. "Bill"	Adv. Flight Instructor	0
Donoghue	Ron	LAC	18
Douglas	W.S.	AV/C; USAAF	14
Dove	W.S.M.	LAC	3
Dowdell	E.H.	AC2	24
Downs	J.H.	LAC	10
Downs	J.T.	LAC	19
Downs	V.L.	PFC; USAAF	0
Dowthwaite	Peter W.	LAC	19
Doyle	P.S.	AC2	22
Dransfield	Don	AC2	20
Drew	R.J.	LAC	19
Drummond	J.F.	LAC	17
Dryden	Katherine	Flight Dispatcher	0
Duck	Stanley W.M.	LAC	16
Duddy	P.G.	LAC	16
Duffey	G.	LAC	5
Dugdale	H.	AC2	21
Duguid	Eric	LAC	19
Duncan	A.H.	LAC	17
Duncan	Hank	Adv. Flight Instructor	0
Durbridge	Keith W.	AC2	23
Durrant	B.F.	LAC	9
Dyer	C.E.	AC2	22
Eales	W.J.	LAC	16
Eames	R.	AC2	21
Easter	C.R.	AC2	23
Easter	Stanley H.	LAC	13
Easton	R.	AC2	21
Eaton	C.W.	LAC	13
Eaton	D.R. "Doug"	LAC	20
Ecclestone	R.V.	LAC	12
Ecclestone	T.S.	LAC	3
Edard	D.W.H.	AC2	23
Edbrooke	E.C.	AC2	24
Eden	S.G.	LAC	16
Edge	R.W.	LAC	3
Edgington	Anthony N.	LAC	15
Edmonds	J.	AC2	23
Edser	R.E.	LAC	8
Edwards	Betsy	Link Trainer Instructor	0
Edwards	G.A.	AC2	22
Edwards	J.L.	LAC	8
Edwards	Percy M.	LAC	19
Edwards	W.S.	LAC	2
Egan	G.H.	AC2	21
Eggensperger	Kermit A.	AV/C; USAAF	19
Elder	F.H.	LAC	3
Elkerton	A.H.	LAC	16
Elkington	G.F.	LAC	9
Elliott	W.E.	LAC	16
Ellis	S. Roye	AC2	20
Ellison	David J.	LAC	20
Ellwood	J. Kenneth	LAC	17
Elverson	R.A.	LAC	10
Emery	Peter	Sir	27
Emmerson	Ron J. B.	AC2	20
Emmett	Dennis R.J.		26
Errington	D.	LAC	19
Evans	Douglas J.	LAC	17
Evans	E.A.J.	LAC	13
Evans	E.J.	LAC	10
Evans	G.W.	LAC	9
Evans	J.C.	LAC	13
Evans	R.W.	LAC	4
Evans	T.R.	LAC	17
Evans	W.	LAC	16
Everard	E.C.	LAC	25
Facer	John		27
Faguy-Cote	A.F.	AV/C; USAAF	17
Farman	Roy E.	AC2	23
Farnsworth	Saunders P.A.	LAC	16
Farrer	W.R.	LAC	12
Farrington	E.	LAC	17
Faulkner	J.F.	LAC	5
Fawcett	Stan	AC2	25
Fearns	R.D.	LAC	6
Featherstone	P.D.	LAC	12
Felton	J.W.H.	AC2	22
Fennell	J.A.	LAC	19
Ferris	D.W.	AV/C; USAAF	14
Fey	B.	LAC	5
Fichter	Virgil C.	Adv. Flight Inst. Cdr.	0
Fidgin	L.P.	LAC	11
Finch	R.C.	LAC	19
Finch	Ken W.B.	AC2	23
Findlay	Alex		26
Findlay	H.A.	LAC	15
Firth	E.	LAC	10
Fisher	J.	LAC	6
Fisher	M.E.	LAC	19
Fisher	William M. "Bill"	Link Trainer Instructor	0
Fitzgerald	M.J.	LAC	6
Fixter	J.	LAC	18
Flahey	Norman A.	AC2	22
Flannigan	G.	LAC	8
Flatman	M.B.	LAC	13
Flawn	R.G.	LAC	14
Fleming	William	LAC	19
Flemons	E.L.	LAC	12
Fletcher	J.	LAC	14
Fletcher	J.R.	LAC	12
Flournoy	R.W.	AV/C; USAAF	19
Ford	A.	LAC	17
Ford	D.G.	LAC	10
Ford	Marc	Ground School Instructor	0
Fordham	Jack B.	Eliminated	9
Fordham	John "F.?"	LAC	9
Forester	W.R.	LAC	4
Foster	J.	LAC	16
Foster	R.S.	LAC	16
Foster-Ogg	W.W.	AC2	23
Fowler	H.	AC2	22
Fowler	W.M.	AV/C; USAAF	14
Fox	P.G.	LAC	16
Foxell	J.S.	AC2	21
Foy	T.	LAC	17
Foy	William	LAC	19
Francis	E.H.	LAC	1
Franklin	E.G.	AC2	21
Frazer	Ceylon B.	Flight Instructor	0
Frazer	Feller	Unknown Position	0
Frazier	Don C.	Flight Instructor	0
Fredericks	Jim		26
Freeman	Dumas	Link Trainer Instructor	0
Freeman	J.W.	LAC	18
French	B.M.	LAC	21
French	E.J.	AC2	21
Frere	Kenneth H.B.	LAC	12
Friedman	I.L.	AV/C; USAAF	19
Friggens	J.S.	LAC	19
Frith	B.R.	AC2	23
Frizell	C.G.	Fl. Lt.	0
Frodey	Ray C.	AV/C; USAAF	18
Frost	E.	AC2	22
Fruen	F.	LAC	20
Fry	P.W.	LAC	9
Fryer	K.	AC2	25
Fulton	R.J.	AC2	24
Funk	Willis L.	Flight Dispatcher	0
Furrier	John D.	Staff	0
Fussell	M.A.	LAC	3
Fyfe	S.	AC2	22
Gabbey	J.	LAC	10
Gale	R.	AC2	23
Gardner			26
Gardner	J.H.G.	AC2	25
Gardner	P.W.L.	LAC	18
Garside	F.	LAC	4
Gartshore	Edwin Boyd	Fl. Lt.; RCAF Instructor	0
Gates	W.T.	AV/C; USAAF	15
Gauer	Willard	Ground School Instructor	0
Gault	S.A.	LAC	15
Gausden	Rodney T.	LAC	12
Geary	N.F.	Squadron Leader	26
Gentry	Jay L.	Flight Instructor	0
Gibbin	L.C.	LAC	2
Gibbon	J.	LAC	13
Gibbons	J.R.	Group Captain	12
Gibbons	M.G.	LAC	13
Gibbs	D.U.	LAC	8
Gibbs	G.	Cpl.; USAAF	0
Gibson	A.	LAC	5
Gibson	Della	Parts Deptartment	0
Gibson	E.	LAC	2
Gibson	Guy E.	AV/C; USAAF	19
Gibson	H.W.	LAC	17
Gibson	J.	AC2	21
Gibson	R.D.B.	LAC	18
Giddings	Bill	LAC	19
Gifford	G.E.	F/SGT	14
Gilbert	D.A.	AC2	25
Gilbert	G.S.W.	LAC	20
Gilbert	Marjorie	Flight Dispatcher	0
Giles	R.C.	AC2	25
Gill	P.M.	AC2	23
Gilroy	R.A.	LAC	2
Gilroy	R.A.	LAC	1
Ginnis			27
Glover	F.J.	LAC	18
Glover	J.	LAC	5
Glover	M.J.	LAC	1
Glover	T.E.	AC2	21
Goatman	Leonard J.W.	AC2	24
Godfrey	E. Norman	LAC	2
Godfrey	Dennis A.	LAC	16
Godwin	G.C.	AC2	22
Gold	C.W.	AC2	20
Goldfinch	J.A.	AC2	25
Golding	G.T.	AC2	20
Goldman	A.	LAC	18
Gomm	J.H.	AC2	22
Gooch	P.E.	LAC	13
Goodchild	E.A.	LAC	14
Goodwin	Fred D. "W."	Flight Instructor	0
Goodwin	K.	LAC	11
Gordon	J.M.	AV/C; USAAF	15
Gorsuch	B.H.	LAC	18
Gossman	Millard C.	Flight Instructor	0
Gough	S.H.	LAC	1
Gould	G.R.	AV/C; USAAF	19
Grace	Paul B.	Ground School Instructor	0
Graco	Paul B.	Aircraft Maintenance	0
Graham	H.J.	LAC	16
Graham	James H.	Flight Instructor	0
Graham	Jimmy S.	LAC	6
Graham	W.J.	LAC	10
Grange	F.L.	LAC	25
Grant	B.	LAC	17
Grant	G.J.C.	LAC	17
Grantham-Stones	R.	AC2	25
Gravell	J.R.	AC2	22
Gravestock	J.H.	AC2	24
Gray	C.K.	AC2	23
Gray	N.	AC2	25
Gray	Peter	LAC	19
Gray	Philip E.	AC2	20
Greasley	J.	LAC	13
Green	Billye Ann	Asst. Stenographer	0
Green	Coy Gilbert	Aircraft Maintenance	0
Green	D.	AC2	24

LAST NAME	FIRST NAME	RANK	COURSE NO.
Green	David J.	Ground Captain	1
Green	G.	LAC	9
Green	G.N.	LAC	2
Green	R.M.	AV/C; USAAF	19
Green	Ron J.	LAC	20
Greens	R.C.	AC2	22
Greenwell	L.	LAC	2
Greenwood	R.P.	LAC	14
Gregory	G.S.	LAC	18
Greives	G.W.	AC2	21
Gresty	T.E.	AC2	20
Greville	W.P.L.	LAC	16
Gridley	R.F.	LAC	5
Grieve	Adam	LAC	18
Griffin	D.J.	AC2	23
Griffiths	F.H.	LAC	4
Griffiths	H.H.	AC2	25
Griffiths	John R.	LAC	6
Griffiths	N.	LAC	14
Grigg	G.	LAC	16
Grigsby	Doyle E.	Adv. Flight Instructor	0
Grimshaw	P.B.	LAC	13
Grogan	C.D. Paddy	LAC	16
Grove	Ivor D.	AV/C; USAAF	19
Groves	R.E.	LAC	5
Guy	Gordon D.	AC2	25
Gynn	S.T.	AC2	25
Hacker	Gordon	LAC	15
Hackforth	P.J.	LAC	16
Haggan	David A.	Rev.	27
Haggett	D.W.	AC2	22
Hague	J.	AC2	22
Haig	J.S.	LAC	19
Hale	I.E.	AC2	23
Hall	B.H.	LAC	3
Hall	J.A.	LAC	3
Hall	J.H.	LAC	13
Hall	P.	LAC	2
Hall	P.	LAC	3
Hall	W.A.L.	LAC	13
Hall	W.C.	LAC	8
Hallam	W.C.A.	LAC	19
Hallett	H.V.D.	AC2	22
Halligan	Douglas A.	CPL	17
Hallowell	F.G.	LAC	17
Halstead	K.	LAC	4
Halvorsen	Gail S.	AV/C; USAAF	19
Hamer	J.G.	AC2	22
Hames	C.S.	CPL	10
Hammerton	W.C.	LAC	16
Hammitt	Haldon J.	Adv. Flight Instructor	0
Hammond	A.D.	LAC	3
Hammond	H.	LAC	16
Hammond	Reg M.	AC2	22
Hampton	Fred D.	AC2	21
Handley	C.	Fl. Lt.	0
Hands	R.L.	LAC	10
Hankin	P.	AC2	22
Hanson	J.G.	LAC	16
Haralson	John	Flight Instructor	0
Harben	R.M.	LAC	16
Harbottle	W.H.J.	LAC	6
Harcourt	J.W.G.	LAC	16
Hardie	D.W.	LAC	18
Harding	H.E.	AC2	20
Hardman	M.D.	LAC	17
Hardy	G.D.	LAC	16
Hare	T.R.	AC2	25
Hargreaves	T.	LAC	10
Haries	C.	LAC	4
Harker	T.E.B.	LAC	15
Harland	S.T.	LAC	13
Harman			26
Harms	Peter M.	LAC	18
Harper	Frank Ralph	AC2	21
Harrad	John H.	LAC	1
Harris	Alan F.	AC2	25
Harris	C.A.	LAC	13
Harries	G.J.	AC2	20
Harris	J.	LAC	10
Harris	James E.	Flight Instructor	0
Harris	R.B.	AV/C; USAAF	16
Harris	W.R.	LAC	16
Harrison	Chris E.	LAC	13
Harrison	G.C.	LAC	6
Harrop	E.	LAC	14
Harry	C.	AC2	23
Hartill	J.E.A.	LAC	3
Hartley	D.H.	AC2	21
Harvey	K.A.	LAC	15
Harvie	C.	AC2	25
Haslam	Brian	LAC	13
Hastings	D.C.	AC2	25
Hatton	F.G.	AC2	24
Haw	C.L.	LAC	13
Hawkes	E.O.	AC2	24
Hay	C.M.	LAC	15
Hay	H.	LAC	6
Hayes	John Newton	Flight Instructor	0
Haylock	G.S.	AC2	22
Haynes	R.S. Bob	LAC	19
Haynes	C.	LAC	9
Head	W.	AC2	22
Heath	Alan F.	LAC	5
Heath	C.G.W.	AC2	23
Heath	R.	AV/C; USAAF	19
Heaton	Paul R.	AC2	23
Hedinger	J.F.	LAC	17
Heeley	P.J.	LAC	3
Heinmiller	Ben	Unknown Position	0
Helling	G.H.	LAC	19
Hellyar	J.E.	LAC	15
Heminger	Edward Clarence	Adv. Flight Instructor	0
Hemmings	F.	LAC	5
Henderson	D.	LAC	16
Henderson	Gordon	LAC	7
Hendrick		W/O; USAAF	0
Hendrix	Gail	Staff	0
Henley	E.C.	AC2	20
Henley	W.E.	LAC	17
Henson	N.J.	AC2	20
Herbert	A.F.	CPL	11
Herbert	A.J.	AC2	24
Herbert	Stanley	Ground School Staff	0
Herndon	Ken(m) Edwin	Flight Instructor	0
Hesketh	E.	LAC	15
Hetherington	T.	LAC	6
Hetzel	Robert A.	Flight Instructor	0
Hewat	H.J.	CPL	16
Hewitt	Cyril M.	AC2	23
Hichol	C.W.	LAC	9
Hick	J.L.	LAC	10
Hick	R.H.N.	SGT	13
Higgins	G.	LAC	17
Higgs	G.	LAC	9
Higham	B.F.	LAC	9
Hill	B.	LAC	14
Hill	Charles O.	Flight Instructor	0
Hill	E.H.L.	LAC	18
Hill	G.S.	LAC	16
Hill	H.S.O.	AC2	24
Hill	H.T.G.	LAC	13
Hill	S.	AC2	21
Hillard	Carl	Flight Instructor	0
Hillard	R.E.	AC2	21
Hillis	John D.	AV/C; USAAF	16
Hillman	William	Flight Instructor	0
Hills	Brian	Wing Commander	26
Hills	B.C.J.	LAC	22
Hilton	F.W.	Wing Commander	0
Hilton	John A.	Adv. Flight Instructor	0
Himes	Winfield S.	AV/C; USAAF	17
Hindmarsh	A.D.	Squadron Leader	12
Hiscox	A.	AC2	23
Hiscock	D.C.	AC2	25
Hiscock	R.A.	AC2	24
Hiscutt	R.C.	AC2	24
Hitchcock	P.F.	AC2	22
Hitchman	Claude D.	LAC	18
Hoare	Don H.	LAC	6
Hoccom	D.C.	LAC	10
Hocking	D.R.	AV/C; USAAF	17
Hodkins	R.D.	LAC	4
Holden-Hindley		LAC	14
Holdstock	Len F.J.	Cpt.	5
Holdsworth	W. "Bob"	LAC	20
Holland	J.A.	LAC	18
Hollenbeck	Miss Leatrice	Staff	0
Holliday	R.J.S.	AC2	21
Holloman	Leona M.	Spartan Newsletter	0
Holloway	E.J.	LAC	2
Holmes	Bernold Turi "Bernie"	Flight Instructor	0
Holmes	D.	AC2	23
Holmes	N.L.	AC2	23
Holmes	P.	AC2	20
Holoman	Leona	Spartan Newsletter	0
Homer	D.M.	AV/C; USAAF	19
Hooten	A.D.	AC2	21
Hope	A.	LAC	7
Hope	A.	LAC	16
Hopkins	Henry W.	Control Tower	0
Horney	N.E.	LAC	15
Horrex	J.H.	LAC	17
Horsfield	H.	LAC	16
Horsley	Hugh	AC2	24
Horwood	A.J.	LAC	1
Houedard	C.L.	LAC	4
Houghton	K.W.	LAC	9
Houghton	W.	LAC	10
Howard	I.F.	AC2	22
Howarth	J.K.	LAC	2
Howarth	J.L.	AC2	21
Howell	W.A. "Bill"	AC2	20
Howells	J.	LAC	1
Hoyle	E.	LAC	9
Hudson	Thomas J.	AC2	20
Hudson	W.A.	LAC	10
Hudson	William	Aircraft Mechanic	0
Huggett	Aubrey G.	LAC	19
Hughes			26
Hughes	Elmer E. "Skip"	Link Trainer Instructor	0
Hughes	P.A.	LAC	2
Hugo	P.J.	LAC	4
Hull	Cy	Staff	0
Hull	G.F.	LAC	8
Hulme	R.C.	AC2	25
Hulyer	Arthur F.	LAC	17
Humphries	P.L.	LAC	13
Hunt	A.S.	LAC	8
Hunt	C.S.		26
Hunter	A.N.	LAC	3
Hunter-Blair	D.G.	LAC	1
Hurry	Cyril	LAC	23
Hurst	A.H.	LAC	13
Hurst	G.E.	AC2	21
Hurwitz	Monroe S.	Flight Instructor	0
Hussey	Roy J.H.	LAC	3
Hutchings	K.A.	AC2	22
Hutchins	A.V.S.	LAC	16
Hutchinson	H.	AC2	23
Hutchinson	Jim	LAC	13
Hutt	J.A.	AV/C; USAAF	18
Hutton	K.T.	LAC	10
Hyslop	J.O.	LAC	12
Ibbett	Michael J.	LAC	18

LAST NAME	FIRST NAME	RANK	COURSE NO.
Ingersoll	Winthrop	Flight Instructor	0
Ingham	T.	LAC	18
Inglas	A.F.	AC2	22
Ingle	Juanita	Link Trainer Instructor	0
Iniff	H.R.N.	LAC	16
Inman	Donald E.	AV/C; USAAF	16
Ireland	D.B.	LAC	19
Irish	A. James	AC2	25
Irving	E.H.	AV/C; USAAF	18
Jack	J.D.	AC2	23
Jacklin	D.W.G.	AC2	24
Jackson	B.	LAC	18
Jackson	G.O.	AC2	20
Jackson	J.S.	AC2	23
Jackson	K.	LAC	10
Jackson	Michael	AC2	20
Jackson	P.L.G.	AC2	24
Jackson-Moore	Roy		26
Jacobs	Harold A.	AV/C; USAAF	16
Jacobs	L.J.	AC2	24
James	B.	AC2	22
James	Jack	Link Trainer Instructor	0
James	W.D.	LAC	16
Jameson	Richard Marlow	Adv. Flight Instructor	0
Jamison	D.H.	LAC	18
Jardine	B.R.	LAC	12
Jarrett	E.C.	AC2	21
Jarrett	F.H.	LAC	17
Jarvis	C.A.	AC2	20
Jaten	C.H.	AV/C; USAAF	13
Jefferies	J.O.	LAC	16
Jeffrey	F.	AC2	22
Jeffrey	T.	AC2	25
Jenkin	A.T.	AC2	23
Jenkin	B.	LAC	12
Jenkins	F.H.	LAC	4
Jenkins	Glen T.	AC2	23
Jenkins	J.G.	LAC	6
Jenkins	O.C.T.	LAC	12
Jepson	G.	LAC	18
Jess	Ron E.	LAC	13
Johnson	B.M.	USAAF	0
Johnson	D.	AC2	21
Johnson	Donald D.	AV/C; USAAF	15
Johnson	H.F.	LAC	1
Johnson	Harry	LAC	17
Johnson	Ken W.	Squadron Leader	23
Johnson	Phillip A.	LAC	19
Johnson	Pauline	Staff	0
Johnson	R.J.	SGT	7
Johnson	Raymond C.	Flight Instructor	0
Johnston	D.	AC2	25
Johnston	Edith	Hangar 1 Staff	0
Jones	B.H.	AV/C; USAAF	16
Jones	B.L.	AC2	21
Jones	C.	LAC	12
Jones	D.	LAC	10
Jones	E.W.	LAC	3
Jones	G.A.	AC2	23
Jones	G.C.D.	AC2	21
Jones	H.	LAC	12
Jones	I.	LAC	13
Jones	I.A.	LAC	4
Jones	I.B.	LAC	6
Jones	J.L.	LAC	1
Jones	J.R.	LAC	4
Jones	L.	AC2	21
Jones	M.	Wing Commander	3
Jones	Mary	Staff	0
Jones	R.H.	LAC	8
Jones	Ralph	LAC	7
Jones	Thomas O.	LAC	12/13
Jones	W.A.	LAC	5
Jordan	Victor C.	AC2	22
Jordan	W.	LAC	10

LAST NAME	FIRST NAME	RANK	COURSE NO.
Jull	S.J.	LAC	14
Jupp	P.A.	LAC	4
Justin	S.H.	LAC	16
Kauff	James E.	Flight Dispatcher	0
Kaye	B.J.G.	LAC	4
Keeling	J.A.	LAC	4
Keen	Frank M.	LAC	18
Keily	J.H.B.	LAC	10
Keith	Walter	Staff	0
Kell	J.A.	LAC	15
Kelly	D.T.	LAC	6
Kelly	L. Mcl.	US	8
Kelly	Terry A.		26
Kennedy	John	LAC	18
Kennington	George	LAC	12
Kent	G.E.	LAC	12
Kent	V.E.S.	AC2	21
Kenwright	B.R.	LAC	4
Kenyon	Alan	LAC	1
Kenyon	T.	LAC	4
Kerby	R.S.	LAC	3
Kermode	A.C.	Wing Commander	0
Kerr	F.	AC2	23
Kershaw	E.	LAC	19
Kesler	H.B.	AV/C; USAAF	17
Kevern	W.H.	AC2	22
Khan	M.D.	LAC	3
Kidd	A.	LAC	16
Kieffer	J.M.	AV/C; USAAF	18
Kiley	B.J., Jr.	AV/C; USAAF	15
Kimmond	A.F.	AC2	25
Kind	Leslie V.	AC2	25
King	D.C.	AC2	22
King	D.J.	LAC	1
King	D.N.	LAC	14
King	R.S.	LAC	3
King	T.F.	SGT	14
King	W. Alex L.	LAC	19
Kinman	J.	LAC	13
Kinmond	"Jock" Russell	Cadet	0
Kiple	F.K.	Chief Gr. School Inst.	0
Kirchner	Harry M.	Staff	0
Klein	K.J.	AC2	23
Klobacha	John J.	AV/C; USAAF	15
Knee	R.F.	AC2	25
Knight	R.	LAC	11
Knott	C.E.	LAC	5
Koepenick	Fred G.	Flight Instructor	0
Kotteles	George J.	Flight Instructor	0
Krucker	Bob	Stenographer	0
Kukula	Ted	Staff	0
Lachenmaier	Robert W.	Flight Instructor	0
Ladd	J.O.	LAC	9
Ladley	I.	LAC	11
LaForce	P.R.	AV/C; USAAF	15
Lake	W.W.	AV/C; USAAF	19
Lakin	B.G.	LAC	13
Lamb	Ronnie O.	LAC	3
Lancaster	Ivan	Squadrn Leader	19
Lander	F.T.	AC2	24
Lane	Orville	Field Dispatcher	0
Lane, Jr.	William L.	Link Trainer Instructor	0
Lang	Don	LAC	14
Langer	J.F.	A/Cdr.	17
Langford	G.A.	LAC	12
Langtry	Reg V.	Cpt.	14
Lansford	Theodore (Thomas) C.	Adv.Flight Instructor	0
Laraway	D.W.	AV/C; USAAF	14
Lasiter	Carl	Aircraft Maintenance	0
Latta	Pauline	Staff	0
Latta	William Mitchell	Flight Instructor	0
Laundy	S.C.J.	LAC	5
Lawrance	A.A.	LAC	17
Lawrence	N.H.	LAC	18
Layzell	A.E.	LAC	14

LAST NAME	FIRST NAME	RANK	COURSE NO.
Leach	A.	LAC	10
Leadbetter	G.W.	AC2	20
Leader			26
Leather	Colin	LAC	18
Leather	J.G.	LAC	2
LeBrocq	L.S.	LAC	14
Ledbetter	J.E.	Aircraft Mechanic	0
Lee	D.H.	Wing Commander	0
Lee	H.L.	LAC	16
Leigh	G. Keith	AC2	21
Leigh	J.R.	F/SGT	18
Leinaar	Elmer L.	AV/C; USAAF	13
Lemon	William R.	Staff	0
Lennox	W.D.	LAC	19
Lentz	A. Edward	Adv. Flight Instructor	0
Leppert	Newell H.	Adv. Flight Instructor	0
Lerch	Don R.	AV/C; USAAF	19
Leslie	Juanita Lee	Aircraft Maintenance	0
Letsch	Robert D.	Flight Instructor	0
Letsch, Dr.	William	Flight Instructor	0
Levell	T.B.	AC2	22
Lewin	M.	LAC	11
Lewis	G.R.	LAC	6
Lewis	S.	Unknown Position	0
Lewis	T.W.	LAC	8
Lidbetter	P.T.	AC2	21
Lidle	Lois	Staff	0
Lile	Charles H.	Staff	0
Lindsay	George M.	LAC	11
Line	R.	AV/C; USAAF	13
Linehan	C.C.	LAC	9
Lines	Eric H.	LAC	1
Lines	P.	LAC	1
Ling	H.	LAC	11
Lister	H.	LAC	5
Littell	R.M.	AV/C; USAAF	18
Littlefield	Robert Spencer	Adv. Flight Instructor	0
Liversedge	R.S.	LAC	12
Lloyd	F.	LAC	12
Lloyd	H.E.	CPL	14
Lloyd	H.W.J.M.	LAC	4
Lloyd	J.	LAC	15
Lloyd	M.J.	LAC	10
Lloyd-Jones	Bryn	LAC	21
Locke	J.S.	LAC	16
Longbottom	G.	LAC	3
Lord	Jim	LAC	13
Lord	John P.	LAC	5
Lord	R.W.G.	LAC	4
Lotz	George "Red"	Ground School Instructor	0
Lowes	D.E.	AC2	22
Lowry	Mary M.	Asst. Parachute Rigger	0
Lucas	J.D.	LAC	14
Lucas	R.E. "Bud"	Flight Instructor	0
Luxton	J.A.	AC2	22
Lyon	R.	SGT	8
Macauley	A.	AC2	21
MacDonald	Ian R.		27
MacDonald	Lennie P.	LAC	19
MacGregor	Colin	Hon. Sheriff	27
MacGregor	D.S.	LAC	12
MacGregor	W.S.	AC2	23
Machin	D.	LAC	3
Mackanzie	F.R.	LAC	3
MacKay	Angus K.	AC2	25
Mackey	S.P.	AC2	25
MacKinnon	E.C.	LAC	13
Mackworth	R.C.A.	LAC	18
MacLemore	A.B.B.	2nd Lt.; USAAF	0
Macleod	"Roddy"	LAC	23
MacMaster	I.E.	LAC	12
Macnally	John T.	LAC	18
Maddox	Deryk A.	Ground Captain	22
Maddox	Leslie G.		26

LAST NAME	FIRST NAME	RANK	COURSE NO.
Magill	Andrew F.	AV/C; USAAF	18
Mair	M.W.	LAC	14
Major	W.R.	LAC	5
Maley	Robert D.	Aircraft Maintenance	0
Malocsay	George Frank Nelson	Adv. Flight Instructor; Ground School Instructor	0
Malone	Jack L.	LAC	8
Malvern	J.A.B.	LAC	10
Maniex	Larry A.	AV/C; USAAF	16
Mann	W.N. "Jack"	USAAF	17
Manning	L.W.C. "Mick"	AC2	25
Mansell	S.P.A.	LAC	18
Mappin	H.	LAC	6
Marland	C.	AC2	25
Marnie	Jimmie A.	LAC	19
Marr	W.T.	AV/C; USAAF	15
Marsh	A.R.	AC2	21
Marshall	A.W.P.	LAC	16
Marshall	Bert W.	Capt. USAAF	0
Marshall	Colin S.R.	AC2	22
Marshall	K.J.	LAC	5
Marshall	S. John	LAC	19
Marti	H.P.	LAC	12
Martin	F.H.	Fl. Lt., D.F.C.	0
Martin	L.	LAC	12
Martin	L.C.	LAC	5
Martin	R.B.	LAC	4
Marzullo	Raymond J.	AV/C; USAAF	17
Mason	C.	LAC	18
Mason	Don F.	AC2	21
Mason	P.C.	LAC	7
Massey	W.J.	LAC	4
Matheson	A.	LAC	5
Mathews	Derrick	LAC	19
Mattock	A.D.	AC2	22
Maule	J.D.	LAC	15
Maxwell	J.R.	AC2	21
May	W.C.	LAC	14
Maybee	E.L.	LAC	6
Mayer	George	Adv. Flight Instructor	0
Mayo	R.F.	AC2	25
McAra	R.S.	LAC	9
McArthur	H.G.	LAC	22
McCallum	R.L.	AC2	23
McCammon	Chester L.	Aircraft Maintenance	0
McCarthy	B.S.	AC2	22
McCleery	K.D.	AV/C; USAAF	19
McCleod	R.M.M.	CPL	23
McClosky	W.R.	Lt. USAAF	0
McClymont	David	LAC	15
McConnell	Glenn	Adv. Flight Instructor	0
McConough	D.F.	AC2	20
McCormick	John C.	Flight Instructor	0
McCrary	Dee G.	AV/C; USAAF	19
McCrea	J.	AC2	25
McEwan	Ian J.	LAC	19
McFall	A.H.	LAC	2
McGibbon	W.F.	AC2	25
McIntosh	W.	LAC	19
McKelvie	A.M.	LAC	7
McKeon	T.I.	LAC	14
McKeown	David T.	Cpt.	15
McKinney	S.A.	AC2	22
McLaren	A.	LAC	10
McLeob	L.L.	LAC	6
McMillan	H.A.	AC2	20
McMillan	I.R.	LAC	6
McMurdo	Ward H.	Adv. Flight Instructor	0
McNamara	J.F.	LAC	5
McNaugton	Patty	Office Staff	0
McNeill	A.	AC2	20
Mead	H.B.	Lt.; USAAF	0
Meek	Freddie	LAC	19
Mehaffie	Clarence N.	AV/C; USAAF	13
Meiklem	R.C.	LAC	12
Mellodey	W.	LAC	15

LAST NAME	FIRST NAME	RANK	COURSE NO.
Mennell	R.R.N.	LAC	16
Mercer	A.	AC2	20
Merrick	T.W.	AC2	22
Messenger	F.E.W.	AC2	21
Metcalf	J.W.	LAC	11
Metcalf	J.W.	LAC	16
Middleton	J.H.	LAC	10
Milburn	J.	LAC	23
Miles	A.R.	LAC	13
Miles	H.L.	LAC	18
Miller	F.	LAC	11
Miller	Frank	LAC	14
Miller	Ralph	Flight Instructor	0
Millikin	Eugene J.	AV/C; USAAF	16
Mills	D.W.	AC2	24
Mills	Roger F.	LAC	14
Milsom	G.B.	Capt.	13
Milton	A.J.	SGT	7
Minch	G.J.	LAC	15
Ming	Anthony J.	US Dir/3BFTS O.B.E.	0
Miniex	C.A.	AV/C; USAAF	16
Minter	B.E.	AC2	25
Mitchell	D.	AC2	24
Mitchell	Kenneth A.		27
Mocker	W.	LAC	11
Moncrieff	L.	LAC	2
Montague	Francis Michael	Adv. Flight Instructor	0
Moon	B.G.	LAC	15
Moore	R.E.	LAC	14
Moore	W.J.	LAC	2
Moores	R.K.	LAC	15
Morgan	C.H.	LAC	12
Morgan	D.H.	LAC	6
Morgan	F.	AC2	21
Morgan	G.E.	AC2	24
Morgans	Otis J.	Ground School Instructor	0
Morgenstern	Harold H.	Adv. Flight Instructor	0
Morris	A.M.	LAC	19
Morrison	J.D.	LAC; F/O Gunnery Inst.	6
Morrison	W.S.	LAC	11
Morshead	O.H.	LAC	12
Morton	H.D.	AC2	25
Mottram	A.	LAC	3
Mounce	Harry H.	Flight Instructor	0
Muir	J.R.	LAC	12
Mullinax	Rufus Eugene	Flight Instructor	0
Munro	L.	LAC	16
Murdoch	C.J.B.	AC2	25
Murphy	Frederick S.	Flight Instructor	0
Murphy	J.R.	LAC	17
Murray	K.D.	LAC	19
Murton	D.C.	LAC	11
Murton	W.	LAC	17
Mussen	V.A.C.	AC1	25
Mycroft	M.J.	LAC	3
Myers	Robert M.	Prim. Flight Commander	0
Mylam	H.J.	AC2	24
Nann	Robert A.	AV/C; USAAF	19
Nathman	Stephen F.	Flight Instructor	0
Naylor	H.	AC2	21
Neame	G.D.	LAC	14
Neeley	Carl	Staff	0
Neill	T.G.	LAC	12
Neilson	D.	LAC	19
Nelson	R.A.	LAC	15
Ness	B.T.	AC2	24
Newton	R.H.	AC2	21
Nichol	R.J.	AC2	20
Nichols	John R.	Flight Instructor	0
Nichols	R.	AC2	23
Nichols	S.	LAC	18
Nicol	R.	AC2	25
Nightingale	W.	AC2	25
Noble	G.	LAC	16
Noble	L.R.	LAC	15
Norman	D.R.	LAC	16

LAST NAME	FIRST NAME	RANK	COURSE NO.
Norris	B.E.A.	AC2	24
Norris	R.C.	LAC	15
North	J.E.	LAC	10
Norwood	S.	LAC	13
Novak	Henry A.	Sgt. US Army Med. Corp.	0
Oakes	T.G.F.	AC2	25
Oakes	Tommy	Unknown Position	0
Oakey	W.C.(E.)	CPL	8
Oat, Jr.	Frank	Flight Instructor	0
O'Callaghan	B.	LAC	14
Oddy	H.D.	AC2	24
O'Dea	W.P.	AC2	25
Odell	Ken S.	AC2	20
Ogilvie	G.L.	LAC	19
Olerenshaw	H.	LAC	16
Oliver	D.J.	AC2	25
Oliver	William A.	Staff	0
Ollerton	R.	LAC	16
Orchard	D.F.	LAC	15
Orrick	R.	AC2	23
Osmand	P.E.	LAC	16
O'Sullivan	P.D.	LAC	1
Oultram	G.E.	AC2	20
Overmiller	Oran H.	Adv. Flight Instructor	0
Overy	D.F.	AC2	24
Page	L.H.	LAC	2
Page	J.W.	AC2	25
Page-Roberts	J.F.C.	AC2	25
Pain	D.	AC2	24
Parfet	Courtland E.	AV/C; USAAF	13
Parfitt	T.	LAC	2
Park	J.	LAC	12
Park	W.B.	LAC	12
Parker	G.R.	LAC	19
Parkin	K.F.	LAC	16
Parker	S.P.	AC2	21
Parkinson	B.I.	LAC	20
Parkinson	D.C.	AC2	23
Parr	James C.	Pvt. US Army Med. Corp.	0
Parsons	D.R.	AC2	24
Parsons	G.E.	AC2	24
Parsons	J.R.	AC2	21
Partridge	R.A.	LAC	11
Patching	P.J.	LAC	16
Paton	G.A.	AC2	23
Paton	R.A.	Fl. Lt.	0
Payne	A.J.	LAC	1
Payne	H.C.	LAC	3
Payton	Alberta	Ground School Staff	0
Payze	H.H.	LAC	2
Peachey	K.O.	LAC	2
Peake	A.M.	LAC	4
Pearce	D.L.	AC2	25
Pearl	W.W.D.	LAC	4
Pearson	G.M.	LAC	3
Pearson	R.C.	LAC	17
Peaty	J.E.	LAC	17
Peck	D.J.	LAC	16
Pegg	A.	LAC	17
Penney	R.M.	AC2	25
Percy-Smith	C.A.	LAC	17
Perkins	J.C.	AC2	21
Perreault	Francis J.P.	AV/C; USAAF	16
Perry	Claude D.	Link Trainer Instructor	0
Perry	J.F.	LAC	17
Perry	G.V.A.	AC2	22
Persing	Herbert E.	Flight Instructor	0
Peterson	John E.	AV/C; USAAF	16
Peterson	Roland E.	Flight Instructor	0
Petts	S.G.J.	AC2	24
Petty	R.W.	LAC	7
Phillips	E.	LAC	3
Phillips	N.S.	AC2	21
Phillips	Ralph	Sgt.; USAAF Clerk	0
Pickering	J.K.A.	LAC	1

LAST NAME	FIRST NAME	RANK	COURSE NO.
Pickup	G.V.	LAC	18
Pilkington	G.A.	LAC	12
Pillows	Della Mae	Ground School Staff	0
Pinsent	L.A.W.	LAC	12
Piskac	Charles L.	AV/C, USAAF	13
Pitchford	W.D.	SGT	7
Pitt	A.H.	LAC	14
Pittam	E.C.	F. Lt., D.C.M.	0
Pitts	Clarence E.	Staff	0
Pitts	Clifford E.	Flight Instructor	0
Pizzey	L.E.A.	AC2	24
Plant	S.G.	LAC	16
Plimott	A.K.	LAC	17
Ploss	P.J.	AV/C; USAAF	17
Pocock	M.A.	LAC	13
Polan	Russell W.	AV/C; USAAF	16
Pollock	E.A.	LAC	5
Pond	J.L.	AC2	23
Pool	A.	LAC	15
Poole	R.	LAC	19
Popplewell	E.W.	LAC	7
Porter	G.R.	AV/C; USAAF	17
Postemsky	J.V.	AV/C; USAAF	16
Potter	J.D.	AC2	24
Potter	Jane T.	Bookkeeper	0
Pottinger	K.A.	AC2	22
Povey	E.L.	LAC	12
Powell	G.	F/SGT; Adv. Flight Instructor	24
Prescott	I.	LAC	2
Prestwich	Ralph E.	Adv. Flight Instructor	0
Price	R.K.	LAC	12
Price	V.C.	LAC	14
Prigg	T.S.	LAC	3
Pringle	John Thomas	Aircraft Maintenance	0
Prior	R.H.	LAC	1
Pritchard	R.W.	SGT	7
Proctor	Douglas J.	LAC	11
Proud	W.J.F.	AC2	21
Prowles	C.F.	LAC	13
Pryce	R.H.C.	F/O	0
Pullman	R.	LAC	18
Purdey	G.E.	AV/C; USAAF	17
Putnam	William	Link Trainer Instructor	0
Pye	W.	LAC	2
Quine	J.W.	LAC	14
Quisenberry	William Lane	Link Trainer Instructor	0
Ralph	W.C.	LAC	20
Ralston	P.B.	AC2	22
Ramage	W.N.	LAC	2
Rammell	Ted R.	Adv. Flight Instructor	0
Ramsay	J.D.	LAC	1
Ramsden	J.E.	LAC	19
Ramsden	A.	AC2	21
Ramsey	R.L.	AV/C; USAAF	14
Randall	F.W.	AC2	21
Randall	W.E.	LAC	5
Randle	W.S.O.	LAC	1
Range	Arthur R.	AV/C; USAAF	17
Ransome	C.F.	AC2	25
Rawlinson	R.	LAC	18
Ray	D.A.J.	LAC	19
Rayner	E.W.	LAC	13
Rayson	J.W.	LAC	17
Read	P.	AC2	21
Reardon	Robert Jack	Flight Instructor	0
Reay	G.N.	LAC	14
Reece	T.W.	LAC	10
Reed	P.F.	LAC	11
Reeder	D.	LAC	10
Rees	W.G.	LAC	14
Reid	A.	LAC	17
Reid	D.D.	LAC	16
Reid	G.	AC2	22
Reid	R.	LAC	5

LAST NAME	FIRST NAME	RANK	COURSE NO.
Rendle	R.V.	AC2	25
Repko	J.L.	2nd Lt.; USAAF	0
Retz	Charles A.	Flight Instructor	0
Reynolds	Charles H.	Asst. Dir. Administration	0
Reynolds	W.D.	LAC	16
Reynolds	H.	AC2	22
Rhoten	George	Parts Department	0
Rice	Douglas N.	AV/C; USAAF	13
Richards	D.G.	LAC	17
Richards	D.J.	LAC	18
Richardson	C.H.	Squadron Leader; D.F.C.	0
Richardson	G.L.	LAC	16
Richardson	M.G.T.	Sgt.	0
Richardson	Mervyn	Corp.	0
Richey	Kenneth	Staff	0
Richmond	A.B.	AC2	20
Richmond	W.G.	LAC	19
Rickman	Charles	Staff	0
Riddell	C.J.	LAC, KIA 2 May 1945	0
Riddle	M.H.	LAC	2
Ridinger	R.D.	AV/C; USAAF	17
Ridley	O.	AC2	21
Rigrose	W.R.	LAC	4
Rimmer	H.E.	LAC	13
Riner	William L.	Parachute Rigger	0
Roberson	Lawrence E.	Staff	0
Roberts	C.A.	AV/C; USAAF	19
Roberts	Charley Mack	Aircraft Maintenance	0
Roberts	D.V.	AC2	22
Roberts	E.W.	LAC	8
Roberts	L.J.	AC2	24
Roberts	S.	LAC	20
Robertson	R.	CPL	9
Robinette	G.H.	AV/C; USAAF	18
Robinson	A.F.	LAC	14
Robinson	J.	AC2	24
Robinson	John F.	Flight Instructor	0
Robinson	Steve A.	AV/C; USAAF	16
Robinson	T.	LAC	7
Roddy	T.	LAC	13
Roe	K.H.	LAC	15
Rogers	S.	LAC	10
Rogers	R.F.	AC2	21
Romaine	Ralph O.	Flight Instructor	0
Rooke	B.	LAC	8
Rooke	F.C.	AC2	20
Root	Eileen	Link Trainer Instructor	0
Roper	A.J.	LAC	18
Roscow	A.	LAC	17
Rose	R.J.	LAC	17
Roshto	Joe E.	AV/C; USAAF	16
Ross	G.C.	AC2	23
Ross	K.W.	LAC	18
Roth	Frederick D.	US Army Med. Corp. Flight Surgeon	0
Rowe	G.R.	AC2	24
Rowland	A. James T.	Sgt., Ground School-Armament	0
Rowland	R.V.	LAC	3
Rowley	R.A.	AC2	20
Roxburgh	H.A.	Wing Com., A.F.C.	0
Rucker	Bertha Louise	Cafeteria	0
Rudin	W.H.	LAC	18
Ruefer	Kenneth M.	Flight Instructor	0
Rushton	P.L.	LAC	15
Russell	R.F.	AC2	25
Russell-Smith	D.	LAC	15
Russum	K.	LAC	6
Rutherford	R.A.	AC2	22
Rutter	E.H.	AC2	24
Ryan	N.	LAC	17
Ryan	D.	LAC	12
Sadgrove	S.H.	LAC	11
Saft	Wayne Thompson	Adv. Flight Instructor	0
Sage	J.P.	LAC	19

LAST NAME	FIRST NAME	RANK	COURSE NO.
Salisbury	Scott	Flight Instructor	0
Salter	V.F.	LAC	15
Salter	S.W.	AC2	22
Sampson	G.A.J.	AC2	25
Samson	G.H.	AV/C; USAAF	19
Samuels	Harley H." Doc"	Adv. Flight Instructor	0
Sanders	Clarence E.	Flight Instructor	0
Sanders	J.	AC2	20
Sanderson	Warren L.	Adv. Flight Instructor	0
Sapp	Vernon W.	Link Trainer Instructor	0
Sargeant	F.A.	LAC	18
Sassoon	R.J.	LAC	15
Sattin		Corp.	0
Saunders	C.E.	LAC	16
Saunders	D.L.	Adv. Flight Instructor?	8
Saunders	J.K.	LAC	17
Saunders	P.A.	LAC	16
Savako	C.L.	AV/C; USAAF	14
Sawyer	D.C.	LAC	6
Sayer	W.A.	LAC	16
Sayers	K.N.	Wing Comm.; A.F.C.	0
Sayers	R.	AC2	25
Sayfritz	V.H.	LAC	13
Scarlata	Alfred	Chief Flying Instructor	0
Scatchard	E.D.	LAC	17
Schofield	S.	LAC	14
Schroeder, Dr.	Leon	Adv. Flight Instructor	0
Schyter		Unknown Position	0
Scott	F.R.	LAC	18
Scott	I.L.	LAC	15
Sealy-Bell	G.	LAC	17
Seaman	T.L.	LAC	6
Self	A.W.	LAC	4
Sellmeyer	Frank	Flight Instructor	0
Senton	H.	LAC	17
Sergeant	W.	AC2	25
Sewell	B.A.R.	AC2	20
Sexton	D.J.	AC2	22
Seymour	C.D.	LAC	4
Shackleton	A.E.	LAC	1
Sharman	A.J.	LAC	15
Sharman	P.W.	LAC	18
Sharps		F/O	0
Shearman	L.G.	AC2	20
Sheehan	W.T.L.	LAC	1
Sheekey	V.J.	LAC	18
Shelton	W.E.	AV/C; USAAF	15
Shepherd	R.S.	LAC	18
Shepherd	Richard H.	AV/C; USAAF	0
Sherlock	John	Adv. Flight Instructor	0
Sherlow	J.A.	LAC	18
Shields	R.N.	AC2	21
Shipley	Lois	Office Staff	0
Shirkey	J.A.M.	LAC	18
Shook	Arvid K.	AV/C; USAAF	14
Short	G.H.T.	Fl. Lt.	0
Short	K.	F/Sgt.	0
Shorter	G.W.	LAC	17
Shortess	Jane	Ground School Staff	0
Shortess	Robert M.	Link Trainer Instructor	0
Shuyter		Fl. Lt.	0
Sidebotham	G.R.	LAC	11
Sillin	Hugh D.	AV/C; USAAF	16
Sim	T.S.	AC2	21
Simkin	D.P.M.	LAC	14
Simmons	D.H.	LAC	4
Simmons	D.H.	LAC	15
Simpson	John Nolte	Link Trainer Instructor	0
Sims	P.J.	LAC	18
Singer	I.D.	LAC	2
Sjogren	J.K.	AV/C; USAAF	18
Skeldon	A.W.	LAC	18
Skepper	D.L.	LAC	18
Sketchley	P.T.	AC2	22
Sleven	L.G.	LAC	13

LAST NAME	FIRST NAME	RANK	COURSE NO.
Sloan	R.A.	LAC	13
Smalley	N.H.T.	AC2	24
Smartt	J.	Fl. Sgt.; Ground School Instructor	0
Smith	A.A.	AC2	21
Smith	A.H.	LAC	9
Smith	A.N.	LAC	6
Smith	B.	AC2	25
Smith	C.S.	LAC	14
Smith	E. Wilton	Sgt.; Link Tainer Instructor	0
Smith	F.A.	LAC	17
Smith	G.M.	LAC	9
Smith	H.	LAC	9
Smith	Helen	Accountant	0
Smith	J.	Capt.; USAAF	0
Smith	John L.	Fl. Sgt.	0
Smith	K.S.	LAC	18
Smith	K.W.T.	LAC	14
Smith	Leon	Aircraft Maintenance	0
Smith	M.M.S.	AC2	21
Smith	R.	LAC	19
Smith	R.C.	AC2	25
Smith	Richard W.	Chief of Flight Inst.	0
Smith	T.D.	AC2	21
Smith	Thomas	Ground School Instructor	0
Smith	Thomas Richard	Ground School Instructor	0
Smith	W.D.	AV/C; USAAF	16
Smith	W.M.	AC2	22
Smith	W.R.	LAC	3
Smith-Carington	J.H.	LAC	4
Snary	H.R.	LAC	3
Snell	Frederick N.	LAC	19
Softley	F.P.	AC2	20
Somerset	W.C.	AC2	22
Songest	J.J.	LAC	11
Southern	R.	LAC	15
Southeron-Estcourt	E.G.A.	LAC	4
Southey	B.U.	LAC	17
Spain	C.T.	LAC	14
Sparrow	P.H.	LAC	15
Speer	Mary Jane	Staff	0
Spencer	H.A.	AV/C; USAAF	14
Spring	N.D.	LAC	17
Spurlock	H.F.	AV/C; USAAF	14
Staddon	H.J.	AC2	22
Stagg	R.O.	LAC	16
Stanley	E.A.	AC2	24
Stanton	Sydney W.	Adv. Flight Instructor	0
Starnes	Jack	Service Inspector	0
Statham	TA..	AC2	22
Steel	A.C.	LAC	8
Steer	R.	AC2	21
Stephens	C.V.	LAC	6
Stephens	D.	AC2	24
Stephenson	G.	LAC	15
Sterne	M.	AC2	24
Stevens	C.E.	AV/C; USAAF	13
Stevenson	Dwight B.	Staff	0
Stevenson	R.A.	LAC	19
Stewart	James C.	Adv. Flight Instructor	0
Stewart	L.A.	AC2	20
Stickel	E.S.	Adv. Flight Instructor	0
Stocker	D.F.J.	AC2	22
Stone	Louis E.	Ground School Instructor	0
Stores	W.	AC2	20
Storey	J.T.	LAC	13
Story	Charles	Staff	0
Straker	S.	LAC	2
Strange	Sydney J.	Fl. Sgt., Armament Inst.	0
Stredwick	A.H.	LAC	19
Street	D.O.	LAC	4
Street	G.A.W.	LAC	14
Strong	J.R.	AC2	24
Stuff	W.A.	AC2	22
Sturgess	P.C.	AC2	22
Suddarth	Louis A.	Flight Instructor	0
Sutherland	R.F.	Fl. Lt.	0
Sutton	D.A.	LAC	4
Swider	Ray J.	AV/C; USAAF	16
Swire	W.H.	LAC	6
Symonds	W.J.C.	AC2	21
Syrett	A.W.	LAC	13
Szejda	John C.	US Army Medical Corp.	0
Taggart	G.E.	AV/C; USAAF	14
Taggart	H. Edmund	Fl. Lt.; Admin. Officer	0
Tanswell	G.W.	LAC	13
Tanter	P.W.	LAC	19
Taplin	J.H.	AC2	22
Tappan	R.W.	AV/C; USAAF	15
Tate	F.C.	LAC	14
Tattam	F.J.	AC2	21
Taylor	A.S.	LAC	13
Taylor	C.	AC2	22
Taylor	C.E.	LAC	14
Taylor	C.M.	CPL	16
Taylor	J.	LAC	6
Taylor	J.A.	AC2	21
Taylor	J.H.	LAC	4
Taylor	P.W.	AC2	23
Taylor	Ronalee	Field Dispatcher	0
Taylor	Tommy	Staff	0
Taylor	W.	LAC	18
Telford	D.T.	AC2	22
Temple	Eileen	Spartan Newsletter	0
Temple-Smith		Cadet	0
Tennant	V.A.W.	AC2	25
Thaxter	R.J.	AC2	22
Thirkell	J.A.	AC2	25
Thirkettle	M.H.	LAC	19
Thomas	Carl	Aircraft Maintenance	0
Thomas	R.	LAC	19
Thomas	W.A.L.	LAC	16
Thompson	Elgin Conway	Staff	0
Thomson	E.G.M.	LAC	16
Thomson	F.G.	LAC	1
Thomson	J.	LAC	15
Thomson	J.K.	LAC	6
Thorniley	O.R.	AC2	20
Thorns	A.E.	LAC	19
Thornton	G.F.	LAC	8
Thrift	C.J.	AC2	24
Thurber III	William H.	AV/C; USAAF	13
Thursby	R.A.	LAC	9
Thurston	P.	AC2	23
Thwaites	D.E.	LAC	5
Thyer	E.C.	LAC	4
Tideswell	D.M.	LAC	17
Tidswell	Charles Victor	Flight Instructor	0
Tidy	D.	LAC	2
Tilbury	G.W.	AC2	22
Timmermeister	John	Flight Instructor	0
Todd	R.A.	LAC	9
Tomlinson	J.W.	LAC	16
Toone	J.	CPL	8
Topping	J.W.P.	LAC	18
Torry	J.C.	AC2	20
Tottle	P.	LAC	16
Towell	A.C.	AC2	23
Townsend	G.	CPL	23
Townsend	G.R.	LAC	16
Tranchant	C.	LAC	14
Trawocl	Gordon M.	Link Trainer Instructor	0
Treacher	A.E.	AC2	25
Tregilgas	S.	LAC	15
Trevithick	T.	AC2	21
Trezise	E.J.	LAC	5
Trigg	Neil C.	Staff	0
Tucker	P.H.	AC2	20
Tufft	K.	LAC	1
Turley	S.R.R.	AC2	22
Turner	D.A.R.	LAC	17
Turner	G.T.	AC2	22
Turner	George R.	Aircraft Maintenance	0
Turner	J.B.	LAC	17
Tushingham	A.V.	AC2	21
Tweny	E.J.S.	LAC	19
Twite	C.B.	CPL	8
Twomey	D.D.	LAC	12
Tyler	Harry	Aircraft Maintenance	0
Ullyett	R.H.	LAC	15
Ulrich	M.R.	AV/C; USAAF	14
Unitt	A.J.	LAC	13
Vail	F.W.	AV/C; USAAF	13
Valentine	G.	LAC	8
Van Sickle	Harold	Flight Dispatcher	0
Vanatta	Miss	Flight Dispatcher	0
Varley	D.A.	AV/C; USAAF	14
Vaughan	A.H.	AC2	21
Vehner	J.H.	LAC	8
Venables	A.B.	LAC	11
Venton	E.G.	AC2	22
Vicherman	J.	LAC	18
Vickerstaff	R.D.	AC2	21
Vint	Al J.	Flight Instructor	0
Vreuls	W.E.	AV/C; USAAF	19
Waddell	J.	LAC	16
Wade	D.J.	AC2	24
Wadley	A.D.	LAC	19
Wagamon	Lester	Staff	0
Wagner	Arthur H.	Ground School Instructor	0
Wagner	M. John	Ground School Instructor	0
Wakefield	Kenneth	Staff	0
Walker	D.H.	LAC	16
Walker	Willard R.	Staff	0
Wallace	D.	LAC	11
Wallace	Richard R. "Dick"	Adv. Flight Instructor	0
Wallis	E.	SGT	11
Walsh	F.J.	Chief Flying Instructor	0
Walton	A.J.P.	LAC	1
Walton	P.M.R.	LAC	15
Walton	T.H.	LAC	11
Ward	F.H.	LAC	17
Ward	L.C.	AC2	21
Ward	W. E.(G.?)	LAC	16
Warner	J.A.	AC2	22
Warner	R.G.	LAC	16
Warr	D.H.	LAC	3
Warren	A.M.	AV/C; USAAF	19
Warren	D.K.	AC2	23
Waterhouse	J.C.R.	LAC	1
Watkins	J.P.	LAC	12
Watkins	John A.	Link Trainer Instructor	0
Watkins	W.W.	AV/C; USAAF	19
Watsham	P.J.	AC2	22
Watson	S.J.	LAC	7
Watts	G.T.L.	AC2	20
Weaver	Warren Oakley	Adv. Flight Instructor	0
Webb	K.A.	AC2	20
Webb	Wanda Allene	Stenographer	0
Webber	C.C.	LAC	16
Weekes	J.H.	AC2	20
Weibley	Richard W. "Dick"	Flight Instructor	0
Weller	A.G.	AC2	22
Weller	A.J.	LAC	4
Wells	C.	LAC	17
Wells	C.R.	LAC	8
Wells	E.J.	AC2	22
Wells	J.R.	LAC	19
Welsh	J.S.	LAC	15
Welsh	M.G.	F/O	0
Welsh	William F.	Chief of Flight	0
Wentz		Adv. Flight Instructor	0
Wertz	Clair J.	Adv. Flight Instructor	0
Wessale	John O.	AV/C; USAAF	19

LAST NAME	FIRST NAME	RANK	COURSE NO.
West	R.P.	LAC	7
West	S.G.	AC2	23
West	W.J.	AC2	21
Westfield	R.M.	AC2	20
Weston	B.E.	AC2	20
Westthorpe	G.A.	LAC	1
Wheeler	C.A.	LAC	16
Whiffin	G.L.	AC2	23
Whitaker	Ayton R.	Fl. Lt., Chief of Nav. Inst.	0
Whitaker	K.	AC2	22
White	A.	AC2	21
White	C.K.	LAC	3
White	C.L.	LAC	6
White	L.S.	LAC	8
White	R.S.D.	AC2	21
White	Robert "Bob"	Service Inspector	0
Whitefield	K.W.	SGT	8
Whiteland	R.E.	AC2	24
Whiteley	Alton William	Aircraft Maintenance	0
Whitfield	R.	LAC	7
Whiting	G.E.S.	LAC	16
Whitman	Alfred M.	Adv. Flight Instructor	0
Whittaker	A.J.	AC2	24
Whittle	P.J.	LAC	4
Whomersley	H.	AC2	23
Widgery	J.H.	LAC	19
Wignall	J.W.	LAC	15
Wilcox	R.	? Staff ?	0
Wilkins	L.J.H.	AC2	24
Willbie	R.T.	AC2	21
Williams	Clarence Eugene	Aircraft Maintenance	0
Williams	J.S.	LAC	18
Williams	J.S.M.	LAC	15
Williams	K.D.	LAC	16
Williams	P.B.	AC2	23
Williams	P.C.	LAC	15
Williams	R.O.G.	LAC	10
Williams	W.D.	LAC	18
Willis	T.A.	LAC	12
Willis	W.B.	AV/C; USAAF	18
Wilson	A.	LAC	3
Wilson	C.K.	LAC	16
Wilson	F.J.	AC2	20
Wilson	I.D.	AV/C; USAAF	15
Wilson	J.R.	LAC	14
Wilson	L.F.	LAC	14
Wilson	Robert E.	Aircraft Maint. Chief	0
Wilson	Tom L.	Flight Instructor	0
Wilson	Mrs. U.	Cafeteria Staff	0
Wilson	V.A.	LAC	5
Winfield	M.L.	LAC	13
Wing	Bill	Flight Instructor	0
Wing, Jr.	William W.	Staff	0
Winter	Arthur	Corp.; Clerk in RAF Office	0
Wiseman	C.A.	AV/C; USAAF	19
Wisner	William L.	Flight Instructor	0
Withington	A.C.	LAC	18
Witt	H.A.	LAC	3
Woblomoo	George T.	Adv. Flight Instructor	0
Wolfe	J. Marvin	Flight Instructor	0
Wolstenholme	F.	AC2	20
Wood	Charles M.	Flight Instructor	0
Wood	J.	LAC	8
Wood	K.W.	CPL	22
Wood	M.	LAC	13
Wood	R.I.	LAC	16
Wood	T.W.K.	LAC	19
Wood	W.G.	AC2	24
Woods	C.	AC2	24
Woods	G.E.	LAC	13
Woods	P.R.	LAC	11
Woollacott	R.	LAC	10
Woolnough	D.E.	LAC	10
Worley	John C.	Aircraft Maintenance	0
Wright	A.J.	AC2	24
Wright	E.	LAC	16
Wright	E.H.	AC2	24
Wright	Florence K.	Spartan Newsletter	0
Wright	H.E.	LAC	5
Wright	J.	LAC	3
Wright	P.D.	LAC	4
Wroe	G.M.	LAC	14
Wyatt	J.A.	LAC	15
Wyatt	Jack K.	AV/C; USAAF	19
Wyatt	R.H.	AC2	25
Wyatt	S.J.	LAC	13
Wyver	K.	LAC	2
Yachochack	Theodore	Adv. Flight Instructor	0
Yake	H.S.	AV/C; USAAF	14
Yearley	H.W.	LAC	11
Yeazel	George	Asst. Dir. Office	0
Yokum	Carl	Lt. US; Parts Supply	0
Young	D.B.	AC2	20
Young	G.	AC2	22
Young	George F.	Flight Instructor	0
Youngclaus	Warren C.	Chief Ground School Inst.	0
Zdanowicz	F.	AV/C; USAAF	15
Zilar	Billy	Accountant	0

Members of the June 1944 class of No. 3 BFTS graduates line up to receive their wings. *Courtesy Kay Wessale Schnieber*

Heavily dressed cadets prepare for flying training. The fleece lined garments helped to ward off the cold in the open cockpit planes.
Courtesy Phyllis Mann Wiles

— NOTES —

1 *The Ponca City News*, August 25, 1941, p. 4.

2 *Ibid.*

3 Peter Hepplewhite, *Royal Air Force, My War*, (Hodder Wayland Publishing, London: 2003), p. 4.

4 Chaz Bowyer, *History of the RAF*, (Bison Books, Ltd., Hong Kong: 1977), p. 66.

5 *Ibid.*, p. 68.

6 *Ibid.*, p. 69.

7 *Ibid.*, p. 70.

8 *Ibid.*, p. 73.

9 *Ibid.*, p. 76.

10 *Ibid.*, p. 77.

11 Ronald Wakeman, D.F.C., RAF, *As We Were! Calm before the Storm – 1940*, unpublished document, 2005, p. 1.

12 Chaz Bowyer, *History of the RAF*, (Bison Books, Ltd., Hong Kong: 1977), p. 78.

13 *Ibid.*, p. 80.

14 *Ibid.*, p. 78, 84.

15 Chaz Bowyer, *History of the RAF*, (Bison Books, Ltd., Hong Kong: 1977), p. 93.

16 *Ibid.*, p. 92.

17 Brenda Williams, *Britain in the Blitz*, (Jarrold Publishing, Worwich, England: 2003), p. 4.

18 *Ibid.*, p. 3,4.

19 William K. William to Paula Denson, letter, January 8, 2005, author's files.

20 Juliet Gardiner, *The 1940s House*, (Channel 4 Books, Pan MacMillan, Ltd.: London, 2002), p. 78, 79.

21 Emmy E. Werner, *Through the Eyes of Innocents: Children Witness World War II*, (Westview Press: Boulder, Colorado, 2000), pp. 39-45.

22 Nancy Caldwell Sorel, *The Women Who Wrote the War*, (Arcade Publishers: New York, 1999), p. 95, 109, 110.

23 Emmy E. Werner, *Through the Eyes of Innocents: Children Witness World War II*, (Westview Press: Boulder, Colorado, 2000), pp. 43, 45.

24 Brenda Williams, *Britain in the Blitz*, (Jarrold Publishing: Worwich, England, 2003), p. 3.

25 Emmy E. Werner, *Through the Eyes of Innocents: Children Witness World War II*, (Westview Press: Boulder, Colorado, 2000), p. 39-45.

26 Brenda Williams, *Britain in the Blitz*, (Jarrold Publishing: Worwich, England, 2003), p. 4.

27 Chaz Bowyer, *History of the RAF*, (Bison Books, Ltd., Hong Kong: 1977), p. 92.

28 *The Blitz*, http://www.worldwarIIexraf.co.uk, February 16, 2004, p. 1.

29 Brenda Williams, *Britain in the Blitz*, (Jarrold Publishing: Worwich, England, 2003), p. 5.

30 Juliet Gardiner, *The 1940s House*, (Channel 4 Books, Pan MacMillan, Ltd.: London, 2002), p. 105.

31 *The Blitz*, http://www.worldwarIIexraf.co.uk, February 16, 2004, p. 1.

32 *Ibid.*, p. 2.

33 Brenda Williams, *Britain in the Blitz*, (Jarrold Publishing: Worwich, England, 2003), p. 14.

34 *The Blitz*, http://www.worldwarIIexraf.co.uk, February 16, 2004, p. 2.

35 Albert Sims, interview with author, Enid, Oklahoma, June 16, 2005.

36 *The Blitz*, http://www.worldwarIIexraf.co.uk, February 16, 2004, p. 2.

37 *Ibid.*, p. 3.

38 Brenda Williams, *Britain in the Blitz*, (Jarrold Publishing: Worwich, England, 2003), p. 3.

39 *The Blitz*, http://www.worldwarIIexraf.co.uk, February 16, 2004, p. 4.

40 *Ibid.*, p. 3, 4.

41 *American Military History, 1607-1958*, (ROTC Manual No. 145-20: Headquarters, Department of the Army, July 17, 1959), p. 375.

42 Joseph E. Persico, *Roosevelt's Secret War: FDR and World War II Espionage*, (Random House Trade Paperbacks: New York, 2001), p. 83.

43 *American Military History, 1607-1958*, (ROTC Manual No. 145-20: Headquarters, Department of the Army, July 17, 1959), p. 400.

44 Tom Killebrew, *The Royal Air Force in Texas*, (University of North Texas Press: Denton, Texas, 2003), p. 8.

45 *Ibid.*, p. 8.

46 *Ibid.*, p. 8, 9.

47 Joseph E. Persico, *Roosevelt's Secret War: FDR and World War II Espionage*, (Random House Trade Paperbacks: New York, 2001), p. 83.

48 Hugh Morgan, *By the Seat of Your Pants*, (Newton Publishers: Great Britain, 1990), p. 16.

49 Tom Killebrew, *The Royal Air Force in Texas*, (University of North Texas Press: Denton, Texas, 2003), p. 7.

50 *Ibid.*, p. 6.

51 *The Ponca City News*, May 11, 1944.

52 Hugh Morgan, *By the Seat of Your Pants*, (Newton Publishers: Great Britain, 1990), p. 111.

53 *The Ponca City News*, May 11, 1944.

54 Hugh Morgan, *By the Seat of Your Pants*, (Newton Publishers: Great Britain, 1990), p. 112.

55 *The Ponca City News*, August 25, 1941.

56 *The Ponca City News*, August 25, 1961.

57 Brian Barnes to Paula Denson, e-mail letter, October 28, 2005.

58 *Nominal Roll*, (No. 6 British Flying Training School Ponca City Association 4th Edition: England, October 2002), p. 3.

59 Gordon P. "Dave" Davis to Paula Denson, letter, September 9, 2005.

60 NEO A&M College Library, Miami, Oklahoma.

61 Ronald Wakeman, D.F.C., RAF, *As We Were! Calm before the Storm – 1940*, unpublished manuscript, 2005, p. 2.

62 Bert Ginsberg to Paula Denson, telephone interview, September 13, 2003.

63 *Notes for Your Guidance*, (Royal Air Force Publication No. 55729: England, May, 1941), NEO A&M College Library, Miami, Oklahoma.

64 *Ibid.*

65 *Over There: Instructions for American Servicemen in Britain, 1942*, (Bodleian Library, University of Oxford: Oxford, England, 1994), pp. 1-12.

66 Sarah Barrington Westerheide, interview with author, Stillwater, Oklahoma, February 22, 2004.

67 "Blue Laws", *Oklahoma Law*: section 21-918, (http://www.bubbaworld.com/religion_law.html, 2005), p. 4.

68 "Shopping suggestions for RAF cadets entering New York City during World War II", (U.S. Air Force Club handout: NEO A&M College Library, Miami, Oklahoma, general memorabilia, item 6.

69 Stanley John "Jerry" Eaton to Paula Denson, letter, January 10, 2005, author's files.

70 Albert Sims, *The Life and Memories of Albert Sims*, (self published: Enid, Oklahoma, August, 2003), p. 25.

71 *Ibid.*, p. 26.

72 John Cal Price, interview with author, New Milton, England, October 8, 2003.

73 *The Ponca City News*, July 9, 1941.

74 Bob McComb and Paul Victor, T*he Way it Was*, (Personal recollections of two Darr School of Aeronautics flight staff: Part II, unknown date), pp. 11-15.

75 *The Ponca City News*, July 27, 1941.

76 Curtis M. Penland, "Wings of Victory: The British Boys in Oklahoma", (*Joplin Metropolitan Magazine*: Joplin, Missouri, Vol. 1, No. 9, March 1985), p. 23.

77 *Ibid.*

78 *Ibid.*

79 *Ibid.*

80 *Ibid.*

81 *Ibid.*

82 *Ibid.*

83 Chet Peek, *The Spartan Story*, (Aviation Heritage Books, Inc.: Destin, Florida, 1994), p. 114.

84 *The Ponca City News*, September 14, 1941.

85 *The Ponca City News*, January 18, 1942.

86 Mrs. Harold Cogman, interview with author, September 18, 2005.

87 *The Ponca City News*, August 10, 1941.

88 *The Ponca City News*, August 11, 1941.

89 *Ibid.*

90 *The Ponca City News*, August 20, 1941.

91 *Ibid.*

92 *The Ponca City News*, August 25, 1941.

93 *The Ponca City News*, August 29, 1941.

94 *The Ponca City News*, September 14, 1941.

95 *Ibid.*

96 Hugh Morgan, *By the Seat of Your Pants*, (Newton Publishers: Great Britain, 1990), p. 126.

97 *Ibid.*

98 *Ibid.*

99 *Open Post*, (No. 3 British Flying Training School Magazine: Miami, Oklahoma, Vol. 1, No. 9, November 1942), p. 8.

100 Russ Banham, *Conoco: 125 Years of Energy*, (Greenwich Publishing Group, Inc.: Lyme, Connecticut, 2000), p. 93.

101 *The Ponca City News*, May 25, 1944.

102 *The Ponca City News*, April 11, 1944.

103 Bob McComb and Paul Victor, *The Way it Was*, (Personal recollections of two Darr School of Aeronautics flight staff: Part II, unknown date), p. 15.

104 *Nominal Roll*, (No. 6 British Flying Training School Ponca City Association 4th edition: England, October 2002), p. 3.

105 *The Ponca City News*, August 5, 1941.

106 *The Ponca City News*, September 14, 1941.

107 *The Daily Oklahoman*, November 9, 1941, p. 22.

108 *The Ponca City News*, August 18, 1941.

109 *The Ponca City News*, August 19, 1941.

110 *The Ponca City News*, August 18, 1941.

111 *The Ponca City News*, August 19, 1941.

112 *The Ponca City News*, September 4, 1941.

113 *The Ponca City News*, August 20, 1941.

114 *The Daily Oklahoman*, November 16, 1941, p. 2.

115 *The Ponca City News*, August 27, 1941.

116 Rodger Harris, oral history interview with Albert Sims, (Oklahoma Historical Society: Enid, Oklahoma, December 2, 1998).

117 Peter Greaves to Paula Denson, letter, April 17, 2005.

118 *The Ponca City News*, August 15, 1961.

119 *The Ponca City News*, August 27, 1941.

120 *The Ponca City News*, September 2, 1941.

121 *Ibid.*

122 *The Ponca City News*, August 20, 1941.

123 *The Ponca City News*, August 25, 1961.

124 *The Ponca City News*, August 27, 1941.

125 *The Ponca City News*, August 29, 1941.

126 *The Ponca City News*, September 21, 1941.

127 *The Ponca City News*, August 27, 1941.

128 John Cal Price, interview with author, New Milton, England, October 8, 2003.

129 Ibid.

130 Captain A.M.E. "Bert" Spencer, *My War and Peace in Aviation*, (self published: England, no date), p. 15.

131 *Ibid.*

132 Stewart Ross, *On the Trail of World War II in Britain*, (Franklin Watts Publishing: London, 1999), p. 20.

133 Stanley John "Jerry" Eaton to author, letter, January 10, 2005.

134 Barbara Davis, interview with author, January 4, 2004.

135 Edith Bottrell McCallick, interview with author, June 3, 2005.

136 Barbara Davis, interview with author, January 4, 2004.

137 Morris P. Moore, "A Bit of England in Oklahoma", (*The Daily Oklahoman*: Oklahoma City, Oklahoma, November 16, 1941), p. D3.

138 *Ibid.*

139 Albert Sims, interview with author, Enid, Oklahoma, July 8, 2005.

140 "The Link Trainer", (Kirkpatrick Oklahoma Air Space Museum: Oklahoma City, Oklahoma, unknown date), Oklahoma Archive Room, Ponca City Library.

141 Jim Dawson, *The RAF in Arizona, Falcon Field 1941-1945*, (Stenger-Scott Publishing: Newman, Georgia, 2002), p. 104.

142 *Ibid.*

143 *Ibid.*

144 David Pearce, *Dark Skies and Dead Reckoning*, unpublished manuscript, no date, p. 49.

145 *The Ponca City News*, January 6, 1942.

146 *Ibid.*

147 Author attended event.

148 *Nominal Roll*, (No. 6 British Flying Training School Ponca City Association 4th edition: England, October 2002), p. 38.

149 *Nominal Roll*, (No. 6 British Flying Training School Ponca City Association: England, October 2004), p. 4.

150 *The Ponca City News*, August 25, 1961.

151 G.W. "Jack" Welsch, *Life on Elm Creek*, (unpublished manuscript, 1977) pp. 89-93. Oklahoma Archives Room, Ponca City Library.

152 Dr. C. Thomas Hill, interview with author, April 26, 2005.

153 Robert Kelson, interview with author, September 9, 2005.

154 James E. Ledbetter, interview with author, March 10, 2003.

155 Jean O'Kelley, interview with author, May 7, 2005.

156 *The Ponca City News*, July 22, 1943.

157 James E. Ledbetter, interview with author, March 10, 2003.

158 Jean O'Kelley, interview with author, May 7, 2005.

159 Betty Page St. Clair, interview with author, July 10, 2003.

160 *The Ponca City News*, October 28, 1942.

161 *The Ponca City News*, July 4, 1943.

162 Barbara Davis, interview with author, February 1, 2004.

163 Morris P. Moore, "A Bit of England in Oklahoma", (*The Daily Oklahoman*: Oklahoma City, Oklahoma, November 16, 1941), p. D3.

164 *The Ponca City News*, September 14, 1941.

165 *The Ponca City News*, August 25, 1961.

166 *The Ponca City News*, November 5, 1942.

167 *The Ponca City News*, September 15, 1941.

168 *The Daily Oklahoman*, March 26, 1942.

169 *The Ponca City News*, June 4, 1942.

170 *The Ponca City News*, June 8, 1942.

171 Mrs. David Sealey, interview with author, Mickleover, Derby, England, October 7, 2004.

172 *The Ponca City News*, September 14, 1941.

173 *The Ponca City News*, December 28, 1941.

174 The Ponca City News, October 14, 1941.

175 *The Ponca City News*, April 2, 1942.

176 *The Ponca City News*, April 27, 1942.

177 John Cal Price, interview with author, New Milton, England, October 10, 2004.

178 Jack Bowker and Robert Bowker, interview with author, June 18, 2005.

179 Ibid.

180 Morris P. Moore, "A Bit of England in Oklahoma", (*The Daily Oklahoman*: Oklahoma City, Oklahoma, November 16, 1941), p. D3.

181 *Ibid.*

182 *The Ponca City News*, August 7, 1942.

183 *The Ponca City News*, August 27, 1942.

184 *The Ponca City News*, December 10, 1944.

185 Dr. and Mrs. J.E. Fuller, handbill advertising opening of Ponca City USO Club Grand Opening, October 16, 1943.

186 *The Ponca City News*, April 2, 1945.

187 Dr. and Mrs. J.E. Fuller, certificate of appreciation from the United States to Mrs. W.O. Fuller, March 31, 1945.

188 *The Ponca City News*, April 2, 1945.

189 Ponca City Chapter of the Daughters of the American Revolution scrapbook, February 1942.

190 *The Ponca City News*, December 10, 1944.

191 *The Ponca City News*, April 7, 1942.

192 *The Ponca City News*, September 13, 1942.

193 *Tails Up*, (No. 6 British Flying Training School Magazine: Ponca City, Oklahoma, Vol. 19, January 2000), p. 16.

194 *The Ponca City News*, June 14, 1942.

195 *Tails Up*, (No. 6 British Flying Training School Magazine: Ponca City, Oklahoma, Vol. 19, January 2000), p. 12.

196 "British Flying Training School Training in the United States", (United States Air Force declassified document: Maxwell Air Force Base, Alabama, January-December 1943, declassified November 3, 1959), p. 978.

197 Gail S. Ravitts, "Four Yanks in the Oklahoma RAF", (*Flying Magazine*: January 1996), p. 1.

198 *Ibid.*

199 *The Ponca City News*, September 13, 1942.

200 *Ibid.*

201 *The Ponca City News*, April 9, 1942.

202 *Nominal Roll*, (No. 6 British Flying Training School Association, 4th Edition: England, October 2004), p. 38.

203 *The Daily Oklahoman*, April 23, 1944, p. 12C.

204 *Nominal Roll*, (No. 6 British Flying Training School Association, 4th Edition: England, October 2004), p. 38.

205 "United States Army Air Forces History", (http://www.en.wikipedia.org/wiki/United_States_Army_Air_Forces.html, July 26, 2006), p. 1.

206 Gail S. Ravitts, "Four Yanks in the Oklahoma RAF", (*Flying Magazine*: January 1996), p. 2.

207 Olive Floyd, *Pilot Officer James Neale Thorne, RAF, D.F.C.*, (Anthoensen Press: Portland, Maine, 1956), p. 218.

208 Gail S. Ravitts, "Four Yanks in the Oklahoma RAF", (*Flying Magazine*: January 1996), p. 2.

209 *Ibid.*, p. 6.

210 *Ibid.*, p. 7.

211 Gail S. Ravitts, "The Fearsome Foursome Go Forth", (*Flying Magazine*: unknown date), p. 8.

212 Olive Floyd, *Pilot Officer James Neale Thorne, RAF, D.F.C.*, (Anthoensen Press: Portland, Maine, 1956), p. 218.

213 *Ibid.*

214 *Ibid.*

215 Olive Floyd, *Pilot Officer James Neale Thorne, RAF, D.F.C.*, (Anthoensen Press: Portland, Maine, 1956), p. 227.

216 *Ibid.*, p. 226.

217 *Ibid.*, p. 227.

218 *Ibid.*, p. 218.

219 *Ibid.*, p. 338.

220 Jimmy Lichlyter, "A Yank in the RAF", *Tails Up* (No. 6 British Flying Training School Magazine: Ponca City, Oklahoma, May, 1942, No. 5), p. 14.

221 Stanley John "Jerry" Eaton to Paula Denson, letter, January 10, 2005.

222 Harold A. Jacobs, *Top Stick*, (1st Books Library: Bloomington, Indiana, 2002), p. 93.

223 Tom Killebrew, *The Royal Air Force in Texas*, (University of North Texas Press: Denton, Texas, 2003), p. 115.

224 Hugh Morgan, *By the Seat of Your Pants*, (Newton Publishers: Great Britain, 1990), p. 143.

225 Stanley Ward, *Tales Out of Flying School*, (Oak Hill Books: England, 1996), p. 62, 63.

226 G.S. Puddy to Lillian Taylor, letter, August 31, 1987, Oklahoma Archives Room, Ponca City Library.

227 Phyllis Mann Wiles to Paula Denson, e-mail letter, May 10, 2005.

228 Brian Barnes to Paula Denson, e-mail letter, March 4, 2005.

229 Fred Clarke, interview with author, November 19, 2005.

230 John Cal Price, interview with author, New Milton, England, October 8, 2003.

231 Stanley Ward, *Tales out of Flying School*, (Oak Hill Books: England, 1996), p. 58.

232 Pearce, David, *Dark Skies and Dead Reckoning*, unpublished manuscript, no date, p. 29, 30.

233 *Ibid.*

234 Albert Sims interview with author, Enid, Oklahoma, July 8, 2005.

235 *The Ponca City News*, April 20, 1942.

236 Peter W. Smith inscription containing address written on the back of a photograph contained in Mrs. Alta Dixon scrapbook, Ponca City, Oklahoma, 1942. John Cal Price files.

237 *The Ponca City News*, June 19, 1942.

238 Alta Dixon's Ponca City, Oklahoma guestbook, Friends of Marland's Grand Home RAF collection.

239 Information shared with author by various RAF former cadets, Oxford, England, October 9, 2005.

240 *The Ponca City News*, October 16, 1942.

241 *The Ponca City News*, October 9, 1942.

242 *Tails Up*, (No. 6 British Flying Training School Magazine: Ponca City, Oklahoma, Vol. 1, 10, November 1942), p. 18.

243 David Pearce, *Dark Skies and Dead Reckoning*, unpublished manuscript, no date, p. 27.

244 *The Ponca City News*, December 20, 1941.

245 *The Ponca City News*, April 23, 1942.

246 *Nominal Roll*, (No. 6 British Flying Training School Association, 4th Edition: England, October 2002), p. 24.

247 David Riekie, interview with author, Mickleover, Derby, England, October 7, 2004.

248 John Hays, and Eldon Younce, et al, *Anthony, Kansas: Celebrating 125 Years, Active, Attractive and Alive*, (Straight Forward Technologies: Wichita, Kansas, March, 2004), p. 89.

249 John Cal Price, interview with author, New Milton, England, October 8, 2003.

250 Victor Howe to Lillian Taylor, letter, May 22, 1989, Oklahoma Archives Room, Ponca City Library.

251 Hugh Collins to Lillian Taylor, letter, January 17, 1986, Oklahoma Archives Room, Ponca City Library.

252 Brian Barnes to Paula Denson, letter, January 6, 2006. Letter shared portions of earlier letter from D.L. Murray to Bill Barnes, April 20, 1943, Ponca City, Oklahoma.

253 Brian Barnes to Paula Denson, e-mail letter, October 3, 2005.

254 Toni Gibson, interview with author, March 6, 2005.

255 Vincent H. Twomey to Lillian Taylor, letter, September 13, 1986, Oklahoma Archives Room, Ponca City Library.

256 Ibid.

257 B. Brown, "Keep 'Em Riding", *Tails Up*, (No. 6 British Flying Training School Magazine: Ponca City, Oklahoma, Vol. 1, Christmas 1941), p. 21.

258 *Ibid.*

259 *Tails Up*, (No. 6 British Flying Training School Magazine: Ponca City, Oklahoma, Vol. 2, No. 6, February, 1944), p. 10.

260 *Ibid.*

261 Morris P. Moore, "A Bit of England in Oklahoma", (*The Daily Oklahoman*: Oklahoma City, Oklahoma, November 16, 1941), p. D3.

262 Erma Burleson Carmack, interview with author, March 4, 2004.

263 Erma Burleson Carmack, interview with author, May 27, 2004.

264 Mike Igglesden, interview with author, October 7, 2004.

265 David Pearce, *Dark Skies and Dead Reckoning*, unpublished manuscript, no date, p. 27.

266 *The Ponca City News*, January 5, 1942, p. 2.

267 Russ Banham, *Conoco: 125 Years of Energy*, (Greenwich Publishing Group, Inc.: Lyme, Connecticut, 2000), p. 93.

268 Alan Watson, "Personal Diary of Alan Watson", in possession of Kristin Webb.

269 *Tails Up*, (No. 6 British Flying Training School Magazine: Ponca City, Oklahoma, Vol. 1, No. 9, 1942), p. 32.

270 *The Ponca City News*, June 16, 1942.

271 David Pearce, *Dark Skies and Dead Reckoning*, unpublished manuscript, no date, p. 29.

272 *The Daily Oklahoman*, March 1, 1942, p. 5.

273 *The Ponca City News*, July 2, 1943.

274 Original program, author's files.

275 *Tails Up*, (No. 6 British Flying Training School Magazine: Ponca City, Oklahoma, Vol. 2, No. 4, August 1943), p. 22.

276 Stanley Ward, interview with author, October 7, 2004.

277 Stanley Ward, *Tales out of Flying School*, (Oak Hill Books; England, 2000), p. 46.

278 Bryan Bruce-Gardner to Lillian Taylor, letter, May 11, 1984, Oklahoma Archives Room, Ponca City Library.

279 *The Ponca City News*, April 27, 1942.

280 *The Ponca City News*, November 5, 1942.

281 *The Ponca City News*, October 23, 1942.

282 Unknown English magazine, p. 6, Oklahoma Archives Room, Ponca City Library.

283 Morris P. Moore, "So Long Cousins", (*The Daily Oklahoman*: Oklahoma City, Oklahoma, April 23, 1944), p. C12.

284 *Ibid.*

285 *Ibid.*

286 Sarah Barrington Westerheide, unpublished manuscript, Stillwater, Oklahoma, May 7, 2005.

287 Personal knowledge of author.

288 *The Ponca City News*, August 25, 1961.

289 *The Ponca City News*, November 5, 1989.

290 *The Ponca City News*, undated miscellaneous articles recording various events during WWII, Oklahoma Archives, Ponca City Library.

291 Information and addresses used on many letters sent to the Dr. Charles Hocker family by the young men they entertained during their time in Ponca City, author's files.

292 Ibid.

293 Phyllis Hocker, interview with author, April 19, 2005.

294 Jack Portway to Mrs. Charles Hocker, letter, January 10, 1944. Mrs. Hocker noted on the top of the envelope, "Killed May 1945", author's files.

295 Information and addresses used on many letters sent to the Dr. Charles Hocker family by the young men they entertained during their time in Ponca City, author's files.

296 Dick Bolton to Dr. and Mrs. Charles Hocker, letter, October 19, 1944, author's files.

297 *Open Post*, (No. 3 British Flying Training School Magazine: Miami, Oklahoma, Vol. 1, No. 11, February 1943), p. 10.

298 E. Hirst to Dr. and Mrs. Charles Hocker, letter, October 4, 1944, author's files.

299 Stanley Nabbs to Dr. and Mrs. Charles Hocker, letter, August 7, 1944, author's files.

300 Ibid.

301 Ellen Bolton to Mrs. Charles Hocker, letter, November 24, 1944, author's files.

302 Violet Bolton to Mrs. Charles Hocker, letter, January 11, 1946, author's files.

303 Gladys Olorenshaw to Mrs. Charles Hocker, letter, July 17, 1944, author's files.

304 Gladys Olorenshaw to Mrs. Charles Hocker, letter, February 5, 1945, author's files.

305 David Olorenshaw and Birthe Olorenshaw to Paula Denson, letter, May 1, 2005.

306 Gladys Olorenshaw to Mrs. Charles Hocker, letter, April 20, 1945, author's files.

307 Gladys Olorenshaw to Mrs. Charles Hocker, letter, August 26, 1945, author's files.

308 David Olorenshaw and Birthe Olorenshaw to Paula Denson, letter, May 1, 2005.

309 Ibid.

310 Peter Greaves to Paula Denson, letter, April 17, 2005.

311 Author's personal experience.

312 Ella Brown to Mrs. Charles Hocker, letter, March 10, 1946, author's files.

313 Ella Brown to Mrs. Charles Hocker, letter, after February 1, 1947 (postmark cut from the undated letter), author's files.

314 Private writer to Mrs. Charles Hocker, letter, June 8, 1944, author's files.

315 Private writer to Mrs. Charles Hocker, letter, January 22, 1945, author's files.

316 Private writer to Mrs. Charles Hocker, letter, December 12, 1945, author's files.

317 General information gleaned from many of the letters written to the Dr. Charles Hocker family by RAF cadets, author's files.

318 Bernard Rose to Mrs. Charles Hocker, letter, September 27, 1944, author's files.

319 Bill Bishop to Charlotte Taft Weltge, letter, March 6, 1943.

320 Bill Bishop to Charlotte Taft Weltge, letter with footnote in the Alan Watson Diary copy, March 6, 1985.

321 Ibid.

322 Ibid.

323 Street address of Mr. and Mrs. A.E. Taft provided to author by Kristin Weltge Webb.

324 Charlotte Taft Weltge to Bill Bishop, letter, February 17, 1985, provided by Alan Watson's nephew, Tony Norton.

325 Alan Watson, *Condensed Journal of RAF No. 1438732/150428 Alan Norton Watson* (No. 6 British Flight Training School cadet, Course 11, Ponca City, Oklahoma), provided by Kristin Weltge Webb.

326 Charlotte Taft Weltge to Bill Bishop, letter, February 17, 1985, provided by Alan Watson's nephew, Tony Norton.

327 *Nominal Roll*, (No. 6 BFTS Ponca City Association Magazine, 4th edition: England, October 2002), p. 3, 4.

328 Marjorie Shelton, "Summary of Events Leading to the Building of a Memorial to the crew of Wellington X. HF-465", unpublished document, November, 2004, author's files.

329 Marjorie Shelton to Paula Denson, letter, January 28, 2005.

330 Derek Inskeep to Paula Denson, e-mail letter, March 3, 2005.

331 Marjorie Shelton, "Summary of Events Leading to the Building of a Memorial to the crew of Wellington X. HF-465", unpublished document, November, 2004, author's files.

332 Ibid.

333 Ibid.

334 Ibid.

335 Marjorie Shelton to Paula Denson, letter, January 28, 2005.

336 Anne Shelton, "Remember to Never Forget", unpublished document, Toronto Canada, 2004.

337 Ibid.

338 Ibid.

339 Carol Jane Shelton, "Go to Your Sweet Rest", unpublished poem read at the memorial of Wellington X H.F. 465, October 26, 2003.

340 *The Ponca City News*, July 19, 1943.

341 *The Ponca City News*, January 27, 1942.

342 *The Ponca City News*, unknown date.

343 *The Ponca City News*, April 12, 1942.

344 *The Ponca City News*, October 1, 1942.

345 *The Ponca City News*, January 7, 1944.

346 *Nominal Roll*, (No. 6 BFTS Ponca City Association Magazine, 4th edition: England, October 2002), p. 4.

347 Private writer to the Dr. Charles Hocker family, letter, April 22, 1945, author's files.

348 Jack Barrington, "A History of the Darr School of Aeronautics", unpublished manuscript prepared for North Central Oklahoma Historical Association, 1993, p. 1.

349 *Nominal Roll* (No. 6 BFTS Ponca City Association, 4th edition: England, October 2002), p. 4.

350 *The Ponca City News*, March 9, 1942.

351 Scott Murdoch, *Royal Air Force Training in Oklahoma During World War II*, (http://www.airforcebase.net/aaf/raf_ok.html, 2002), p. 1.

352 Gilbert S. Guinn, "British Aircrew Training in the United States, 1941-1945", (*Air Power History Magazine,* Vol. 42, No. 2, Summer 1995), p. 12.

353 *The Ponca City News*, February 19, 1944.

354 *The Ponca City News*, April 16, 1944.

355 *Ibid.*

356 Cinda Hughes, "You Can't Keep a Rez Dog Down", *American Indian Review*, Issue No. 29, Summer 2001, p. 1.

357 Ivy Coffey, "Orbit" Magazine, *The Daily Oklahoman*, May 28, 1972, p. 10.

358 *The Ponca City News*, April 14, 1944.

359 *The Ponca City News*, October 8, 1944.

360 *Ibid.*

361 *The Ponca City News*, July 29, 1948.

362 *Ibid.*

363 *The Ponca City News*, November 16, 1941.

364 *The Ponca City News*, April 23, 1944.

365 Reg Gilbert to Paula Denson, letter, January 18, 2005.

366 *Tails Up*, (No.6 British Flying Training Association: Ponca City, Oklahoma, No. 16, July 1998), p. 12.

367 Reg Gilbert to Paula Denson, e-mail letter, January 26, 2005.

368 Arthur Such, interview with author, Redditch, England, October 6, 2004.

369 Dr. C. Thomas Hill, interview with author, April 22, 2005.

370 Arthur Such, interview with author, Redditch, England, October 6, 2004.

371 Ibid.

372 Ibid.

373 Ibid.

374 Ibid.

375 Ibid.

376 Ibid.

377 Stanley John "Jerry" Eaton to Paula Denson, e-mail letters, January 10, 2005 and April 6, 2005.

378 Jack Youens to Paula Denson, e-mail letter, April 3, 2004.

379 Information gathered in many of the letters in the Hocker letter collection, author's files.

380 Noel Clark and David Olorenshaw to Paula Denson, e-mail letters, January 12, 2005 and April 17, 2005.

381 *The Ponca City News*, November 5, 1989.

382 *The Ponca City News*, April 3, 2005.

383 Sarah Barrington Westerheide, interview with author, Stillwater, Oklahoma, February 22, 2004.

384 Gilbert S. Guinn, "British Aircrew Training in the United States, 1941-1945", *(Air Power History Magazine*, Vol. 42, No. 2, Summer 1995), p. 15.

385 A.V.M. H. Gill to Paula Denson, letter, February 7, 2005.

386 *Miami Daily News Record*, Vol. 24, No. 11, p. 1, July 13, 1941.

387 *Afton American*, June 19, 1941, unknown page.

388 Deryk Maddox, interview with author, Miami, Oklahoma, June 21, 2005.

389 Alan Thomas, *No. 3 British Flying Training School, Miami, Oklahoma: A Brief History*, (No. 3 British Flying Training School Association: England, July 1991), p. 3.

390 *Ibid.*, p. 22.

391 Curtis M. Penland, "Wings of Victory: The British Boys in Oklahoma", *Joplin Metropolitan Magazine*, Vol. 1, No. 9, March 1985, p. 23.

392 *Tulsa Sunday World*, April 18, 1943.

393 Chet Peek, *The Spartan Story*, (Aviation Heritage Books, Inc.: Destin, Florida, 1994), pp. 106-108.

394 *Ibid.*, pp. 109-110.

395 *The Miami Daily News Record*, June 22, 1941.

396 *Open Post*, (No. 3 British Flying Training School Magazine: Miami, Oklahoma, Vol. 1, No. 1, October 1941), p. 9.

397 Velma Nieberding, "Spartan School", The History of Ottawa County, (Ottawa County Historical Society: Miami, Oklahoma, unknown date), pp. 225-226.

398 *The Miami Daily News Record*, July 13, 1941.

399 *Ibid.*, July 17, 1941.

400 *Ibid.*

401 *The Miami Daily News Record*, July 9, 1941.

402 *Afton American*, July 19, 1941.

403 Deryk Maddox, interview with author and photograph provided, Miami, Oklahoma, July 16, 2005.

404 Ibid.

405 "We are Learning to Fly in the United States of America", *The War Illustrated*, January 30, 1942, p. 455.

406 Cyril Grogan, *My Life: Paddy Grogan*, unpublished manuscript, 2005, p. 27.

407 Deryk Maddox to Paula Denson, e-mail letter, August 25, 2005.

408 Ibid.

409 Deryk Maddox, "Graduation Record, No. 3 BFTS, Miami, Oklahoma".

410 Alan Thomas, *No. 3 British Flying Training School, Miami, Oklahoma: A Brief History*, (No. 3 British Flying Training School Association: England, July 1991), p. 2.

411 *Ibid.*

412 Harold A. Jacobs, *Top Stick*, (1st Books Library: Bloomington, Indiana, 2002), p. 83.

413 *The Grove Sun Daily* Internet, (http://www.grovesun.com/archives/2001_obi/obitr.htm, September 10, 2001), p. 6.

414 "Spartan School of Aeronautics Employee Records", Kirkpatrick Museum collection, Oklahoma City, OK, 2002.

415 Ibid.

416 *The Miami Daily News Record*, July 11, 1941.

417 "Spartan School of Aeronautics Employee Records", Kirkpatrick Museum collection, Oklahoma City, OK.

418 *The Ponca City News*, August 5, 1941.

419 Alan Thomas, *No. 3 British Flying Training School, Miami, Oklahoma: A Brief History*, (No. 3 British Flying Training School Association: England, July 1991), p. 3.

420 *Ibid.*

421 Chet Peek, *The Spartan Story*, (Aviation Heritage Books, Inc.: Destin, Florida, 1994), p. 57.

422 *Flight Operations Book, 3BFTS, Miami, Oklahoma*, NEO A&M College Library, Miami, Oklahoma.

423 Cyril Grogan, *My Life: Paddy Grogan*, unpublished manuscript, 2005, p. 22.

424 *Ibid.*

425 *Ibid.*, p. 24.

426 *The Miami Daily News Record*, July 13, 1941.

427 *Open Post*, (No. 3 British Flying Training School Magazine: Miami, Oklahoma, Series 3, Issue 5, Autumn 1998), p. 2.

428 *Miami Daily News Record*, July 13, 1941.

429 *Ibid.*

430 *Open Post*, (No. 3 British Flying Training School Magazine: Miami, Oklahoma, Vol. 1, No. 6, June 1942, p. 13.

431 Paula Denson to Alberta Hutchings, telephone interview, August 18, 2006.

432 *The Miami Daily News Record*, July 13, 1941.

433 *Open Post*, (No. 3 British Flying Training School Magazine: Miami, Oklahoma, Vol. 1, No. 1, October, 1941), pp. 36-38.

434 Alan Thomas, *No. 3 British Flying Training School, Miami, Oklahoma: A Brief History*, (No. 3 British Flying Training School Association: England, July 1991), p. 4.

435 Scott Thompson, *Silver Wings in Prairie Skies*, (KOTV 6 News Special video celebrating VE-Day: Tulsa, Oklahoma).

436 Pilot Officer J.H. Harrad, RAF, V.R., "Hands Across the Ocean", *Open Post*, (No. 3 British Flying Training School Magazine: Miami, Oklahoma, Vol. 1, No. 10, December, 1942), p. 20.

437 Eric Harrop, *Diary of Eric Harrop*, Hendon Museum: London, England.

438 *Roll of Honour, No. 3 BFTS, Miami, Oklahoma 1941-1945*, NEO A&M College, Miami, Oklahoma.

439 *Open Post*, (No. 3 British Flying Training School Magazine: Miami, Oklahoma, Series 3, Issue 5, Autumn 1998), p. 4.

440 Cyril Grogan, *My Life: Paddy Grogan*, unpublished manuscript, 2005, p. 25.

441 *The Miami Daily News Record*, July 13, 1941.

442 Chet Peek, *The Spartan Story*, (Aviation Heritage Books, Inc.: Destin, Florida, 1994), p. 109.

443 *Open Post*, (No. 3 British Flying Training School Magazine: Miami, Oklahoma, final edition, July 2005), p. 8.

444 Original typewritten menu, NEO A&M College No 3. BFTS archives file, Miami, Oklahoma, general memorabilia, item 7.

445 Air Commander J.F. Langer to Paula Denson, letter, September 10, 2005.

446 *Open Post*, (No. 3 British Flying Training School Magazine: Miami, Oklahoma, Vol. 1, No. 2, December 1941), p. 28.

447 *Open Post*, (No. 3 British Flying Training School Magazine: Miami, Oklahoma, Vol. 2, No. 1, May 1943), p. 20.

448 *Open Post*, (No. 3 British Flying Training School Magazine: Miami, Oklahoma, Vol. 2, No. 6, April 1944), p. 19.

449 *Open Post*, (No. 3 British Flying Training School Magazine: Miami, Oklahoma, Vol. 1, No. 3, January 1942), p. 26.

450 *Flight Operations Book, No. 3 BFTS Miami, Oklahoma*, NEO A&M College Library, Miami, Oklahoma.

451 *Open Post*, (No. 3 British Flying Training School Magazine: Miami, Oklahoma, Vol. 3, No. 1, June 1944), pp. 8-9.

452 David Pearce, *Dark Skies and Dead Reckoning*, unpublished manuscript, no date, p. 27.

453 *Flight Operations Book, 3 BFTS Miami, Oklahoma*, NEO A&M College Library, Miami, Oklahoma.

454 *Open Post*, (No. 3 British Flying Training School Magazine: Miami, Oklahoma, Vol. 2, No. 1, May 1943), p. 21.

455 Harold A. Jacobs, *Top Stick*, (1st-Books Library: Bloomington, Indiana, 2002), p. 92.

456 "Camp Crowder, Neosho, Missouri", (http://www.globalsecurity.org/military/facility/camp_crowder.htm, 2005), p. 1.

457 Eric Harrop, *Diary of Eric Harrop*, Hendon Museum: London, England.

458 *Ibid.*

459 Chet Peek, *The Spartan Story*, (Aviation Heritage Books, Inc.: Destin, Florida, 1994), p. 115.

460 *Open Post*, (No. 3 British Flying Training School Magazine: Miami, Oklahoma, Vol. 3, No. 1, June 1944), p. 5.

461 *Ibid.*

462 *Open Post*, (No. 3 British Flying Training School Magazine: Miami, Oklahoma, Vol. 1, No. 2, December 1941), p. 13.

463 *Ibid.*, p. 6.

464 *Open Post*, (No. 3 British Flying Training School Magazine: Miami, Oklahoma, Vol. 1, No. 5, May 1942), p. 31.

465 *Ibid.*

466 *Open Post*, (No. 3 British Flying Training School Magazine: Miami, Oklahoma, Vol. 1, No. 6, June 1942), p. 18.

467 *Open Post*, (No. 3 British Flying Training School Magazine: Miami, Oklahoma, Vol. 1, No. 8, September 1942), p. 6.

468 *Open Post*, (No. 3 British Flying Training School Magazine: Miami, Oklahoma, Vol. 1, No. 10, December 1942), pp. 9-10.

469 *Open Post*, (No. 3 British Flying Training School Magazine: Miami, Oklahoma, Series 3, Issue 5, Autumn 1998), p. 3.

470 First Presbyterian Church of Miami, Oklahoma bulletin, March 19, 1944, author's files.

471 Ron Emmerson to Paula Denson, letter, June 28, 2005.

472 *Miami News Record* picture and caption, unknown date, located in the scrapbook of Mrs. F.M. Hill, Miami, Oklahoma and housed in NEO A&M College Library, Miami, Oklahoma.

473 *Kansas City Star* photograph with caption, unknown date, located in the scrapbook of Mrs. F.M. Hill, Miami, Oklahoma and housed in the NEO A&M College Library, Miami, Oklahoma.

474 Bernard Bower, "Miami's Service Center", *Open Post*, (No. 3 British Flying Training School Magazine: Miami, Oklahoma, Vol. 3, No. 1, June, 1944), p. 8.

475 W.C. Ralph, "20 Course Goes on Leave", *Open Post*, (No. 3 British Flying Training School Magazine: Miami, Oklahoma, Vol. 3, No. 1, June, 1944), p. 25.

476 *Open Post*, (No. 3 British Flying Training School Magazine: Miami, Oklahoma, Vol. 1, No. 11, February 1943), p.13.

477 *Ibid.*

478 *Open Post*, (No. 3 British Flying Training School Magazine: Miami, Oklahoma, Vol. 2, No. 6, April 1944), p. 17.

479 *Open Post*, (No. 3 British Flying Training School Magazine: Miami, Oklahoma, Vol. 2, No. 5, January 1944), p. 20.

480 *Open Post*, (No. 3 British Flying Training School Magazine: Miami, Oklahoma, Vol. 4, No. 1, January 1945), p. 20.

481 *Flight Operations Book, 3 BFTS, Miami, Oklahoma*, NEO A&M College Library, Miami, Oklahoma.

482 *Ibid.*

483 *Open Post*, (No. 3 British Flying Training School Magazine: Miami, Oklahoma, Vol. 1, No. 3, January 1942), p. 34.

484 *The Ponca City News*, January 11, 1942.

485 Lillian Taylor, interview with author, January 21, 2005.

486 *The Woman's Tribune*, Tulsa, Oklahoma, October 22, 1942, unknown page.

487 *Ibid.*

488 *Ibid.*

489 *Ibid.*

490 John Cal Price to Paula Denson, e-mail letter, September 25, 2004.

491 *Tulsa Sunday World*, June 13, 1943, Section 5, unknown page.

492 *Open Post*, (No. 3 British Flying Training School Magazine: Miami, Oklahoma, Vol. 3, No. 1, June, 1944), pp. 20-21.

493 *Ibid.*

494 Miscellaneous files, No. 3 BFTS collection, NEO A&M College Library, Miami, Oklahoma.

495 Pilot Officer J.H. Harrad, RAF, V.R., "Hands Across the Ocean", *Open Post*, Vol. 1, No. 10, December 1942, p. 20.

496 Margaret Hively, unpublished manuscript, No. 3 BFTS Archives, NEO A&M College Library, Miami, Oklahoma.

497 *Open Post*, (No. 3 British Flying Training School Magazine: Miami, Oklahoma, Series 3, Issue 5, Autumn 1998), p. 2.

498 *Flight Operations Book, 3BFTS, Miami, Oklahoma*, NEO A&M College Library, Miami, Oklahoma.

499 Harold A. Jacobs, *Top Stick*, (1st-Books Library: Bloomington, Indiana, 2002), pp. 87-88.

500 Chet Peek, *The Spartan Story*, (Aviation Heritage Books, Inc.: Destin, Florida, 1994), p. 113.

501 *The Sunday Oklahoman*, July 9, 1989, Section A, p. 20.

502 Chet Peek, *The Spartan Story*, (Aviation Heritage Books, Inc.: Destin, Florida, 1994), p. 113.

503 *Ibid.*

504 Newspaper clipping, No. 3 BFTS Archives, NEO A&M College Library, Miami, Oklahoma, undated.

505 G.A.R. Cemetery inscriptions, Miami, Oklahoma.

506 "Raymond Baxter Obituary", (http://www.Telegraph.co.uk, September 16, 2006), p. 7.

507 J.L. Kominicki, *Stars and Stripes*, (http://www.pstripes.com: European Edition, May 18, 1984), p. 1.

508 Gail Halvorsen, "Berlin", episode 4, (http://www.cnn.com/specials/cold.war/episodes/04/interviews/halvorsen/, December 1995) pp. 1-4.

509 Gail Halvorsen, "Survival", Mobility Forum: interview on the 50th Anniversary of the Berlin Airlift", (http://www.findarticles.com/p/articles/mi_qa3744/is_199807/ai_n8886000, July/Aug. 1998), pp. 1-3.

510 Scott Murdoch, *Royal Air Force Training in Oklahoma During World War II*, (http://www.airforcebase.net/aaf/raf_ok.html, 2002), p. 1.

511 *The Miami Daily News Record*, August 23, 1945.

512 Dixie Evans, *Spartan School of Aeronautics Newsletter*, (Spartan School of Aeronautics: Tulsa, Oklahoma, October 1945), unknown page.

513 Deryk Maddox, "Graduation Record, No. 3 BFTS", Miami, Oklahoma.

514 *Flight Operations Book, 3BFTS, Miami, Oklahoma*, NEO A&M College Library, Miami, Oklahoma.

515 Deryk Maddox, Alan Thomas, et al. "Personal comments during the June, 2005 reunion of 3BFTS, Miami, Oklahoma.

A line of Harvard AT6 aircraft wait for their eager occupants. *Courtesy Deryk Maddox*

Seven Ponca City cadets relax in the backyard of Arthur Taft during a weekend leave. *Courtesy Kristen Weltge Webb*

— INDEX —

A

B

C

D

R

S

T

MUSTANG PUBLIC LIBRARY